Frank Kreith
Catherine B. Wrenn

University of Colorado

THE NUCLEAR IMPACT

A Case Study of the Plowshare Program to Produce Gas by Underground Nuclear Stimulation in the Rocky Mountains

WESTVIEW SPECIAL STUDIES ON TECHNOLOGY, NATURAL RESOURCES AND THE ENVIRONMENT

This study was funded by the Research Applied to National Needs Program of the National Science Foundation under NSF Grant No. GI-28423. However, the analysis, conclusions and recommendations presented in this book do not necessarily reflect the views of the National Science Foundation.

Published 1976 in the United States of America by

Westview Press, Inc.
1898 Flatiron Court
Boulder, Colorado 80301
Frederick A. Praeger, Publisher & Editorial Director

Library of Congress Cataloging in Publication Data

Kreith, Frank.
   The nuclear impact.

   Bibliography: p.
   1. Project Plowshare. 2. Gas, Natural--Rocky Mountains. 3. Nuclear excavation. I. Wrenn, Catherine B., joint author. II. Title.
TK9153.K73   622'.33'85   75-31708
ISBN 0-89158-005-0

Printed in the United States of America

Acknowledgments

This book is based largely on an interdisciplinary case study which was a part of the Research Project "The Interaction between Technology, Law, and Politics," supported under the Research Applied to National Needs (RANN) Program of the National Science Foundation (NSF). We wish to express our appreciation for the financial support given to our study by NSF, as well as for a grant from the Council on Research and Creative Work (CRCW) of the University of Colorado to type the final version of our manuscript. We also appreciate the encouragement given us by NSF and CRCW and are especially grateful for the continued support of Mr. A. Konopka, the NSF project manager.

We are deeply indebted to the faculty members at the University of Colorado who were Co-Directors of the NSF-RANN Project on the Interaction between Technology, Law, and Politics: Professor Ronald E. West of the College of Engineering, Professor Donald Carmichael of the School of Law, and Professor William Winter of the College of Arts and Sciences. We are particularly grateful to Ronald West for his tireless efforts in improving and updating the technical sections of the book.

We also wish to express our appreciation to the students who participated in the research seminars of the Technology, Law, and Politics Project at the University of Colorado. Elizabeth Arkell and David Hunter, Research Assistants on the Project, contributed a great deal of the work for the chapters on the Rio Blanco nuclear stimulation experiment. John Van Royan and Malcolm Ronan were helpful in the analysis of the public opinion poll, and David Arkell contributed legal research and analysis.

We obtained much useful information from personal interviews with Dr. Edward A. Martell of the National Center for

Atmospheric Research, Professor John Buechner of the Department of Political Science, Mr. Peter Metzger of the *Rocky Mountain News*, Mr. Harold Aronsen of CER Geonuclear Corporation, Mr. Miles Reynolds, Jr., of Austral Oil Company, Mr. Paul Dugan of Equity Oil Company, and Mr. Tom Ten Eyck, the former Director of the Colorado Department of Natural Resources. Finally, we could not have gone through the many drafts of the book without the typing and editorial assistance of Kay Urry and Rolla Rieder.

Although the case study on which this book is based was funded by the National Science Foundation under NSF Grant Number GI-28423, the analysis, conclusions, and recommendations presented in the book do not necessarily reflect the views of the National Science Foundation, the University of Colorado, the Council on Research and Creative Work, or of any organization and individual who has contributed to this project. The views, opinions, conclusions, and recommendations in this book are solely ours.

Boulder, Colorado, 1975 C. B. Wrenn
F. Kreith

Preface

It is tempting for those of us who opposed underground nuclear experimentation in western Colorado to claim a victory, even though the Rio Blanco blast, the last and most bitterly fought in the Plowshare series, was detonated as scheduled.

Since then:

> Coloradoans voted to prohibit future nuclear blasts in the state without their prior consent.

> The Atomic Energy Commission has been abolished; its twin and conflicting duties of promoting and regulating nuclear technology have been placed in two separate agencies.

> Congress denied further funding for underground nuclear gas stimulation for fiscal 1976, allowing money only to complete evaluation of the Rio Blanco test.

These are gains, to be sure, especially the long overdue demise of the AEC, the federal government's nuclear promoter. Viewed in the broader context of technological assessment and public policy-making, however, these advances serve best to illustrate how outgunned and truly vulnerable we were.

The essential message of this book is that we are no better off today, more than two years after Rio Blanco. The authors of *The Nuclear Impact* detailed the single-minded commitment to nuclear technology which excluded either valid comparisons with other, safer methods or an objective look at the possible effects of the experiments. Concerned individuals, ad hoc groups, and special interest groups are simply no match for government and industry allied behind such projects as Plowshare.

Denied access to complex technical information, citizens are left to base their case on incomplete data. There is no mechanism for independent analysis of the true environmental, social, and economic costs of new technology. The public is asked to rely on the opinions or bland assurances of those committed to a course of action.

Too frequently, such decisions land in the courts. Cast improperly as arbiters of scientific controversy, the courts have no choice but to decide on the weight of the evidence—which, with only rare exceptions, lies lopsidedly with the proponents. Although the context of the book is concerned with the now-dead Plowshare Program, the basic question at issue is still very much alive: What is the role of the public in such policy decisions? That question is as important today as ever, and perhaps more so. The authors sound a clear and timely warning about the pressures our burgeoning energy problems will exert against prudent use of technology, new or old.

The Nuclear Impact carefully charts the pitfalls that await those who will not settle for official assurances and the testimony of experts who have a decided interest in the projects they defend. For that alone, the book is worthwhile. But the authors do not stop with merely defining the problem; they offer concrete proposals for redressing the imbalance which now exists and for ensuring that the public and its experts have a formal role in policy decisions in this technological age. This book hence becomes a valuable tool for those who care about their world and the assaults which can be made upon it in the name of energy or any other crisis.

Floyd K. Haskell
U.S. Senator

Contents

Acknowledgments	iii
Preface	v
Introduction	xi
1 The Plowshare Program	**1**
Legislative History: The Atomic Energy Commission	2
Legislative Authorization of the Plowshare Program	5
The Plowshare Program: Origins and Organization	7
The Search for Direction	8
Funding of the Plowshare Program	12
The Emphasis on Nuclear Stimulation of Natural Gas	13
2 Natural Gas: The Nuclear Route from Reservoir	
to Pipeline	**21**
The Natural Need	21
Nuclear Stimulation—A Necessity?	27
The Way It's Done: The Technique of Nuclear	
Stimulation	29
Radiation: The Unknown Quantity	37
A Plowshare Payoff? Some General Political	
Considerations	42
3 The Course of Project Gasbuggy	**49**
A New Technology Evaluated	56
Gasbuggy Revisited	68
4 Project Rulison	**73**
The Beginnings	73
Action and Reaction, May to September of 1969	84
The Courts, The Governor, and the People	91
The Rulison Results	103
5 Project Rio Blanco: Its Evolution and Impact	**125**
Preliminary Colorado Reactions: Official and	
Unofficial	127
The AEC as Active Participant	132
The Western Slope Scene	141
The Governor, the Government, and the People	150

Politics and the Bureaucracy: Political Reverberations   159
The Tempo Increases   163

6 Technology and Its Evaluation: What Price Rio Blanco? 169
The Projected Plan   170
The Theory and Practice of Phase 1   176
Oil Shale or Natural Gas: A Continuing Controversy?   183
Again, The Court Scene   190
May 17, 1973: Prelude or Finale?   198
Rio Blanco: An Epilogue   202

7 The Nuclear Impact: Reflections Past and Present   209
Plowshare in Colorado: A Reexamination   212
Technology Assessment as a Guide to Future Action   223

Appendix   231
Index   241

List of Tables and Figures

TABLE

2-1   United States Natural Gas Requirements by Class of
      Service (1968, 1980, 1990)   25
3-1   Gasbuggy Production Comparison Data   60
3-2   Gas Composition from the Gasbuggy Well   63
4-1   Predicted and Actual Cavity Dimensions for Project Rulison 105
4-2   Predicted Radiation Dosages to the Local Population from
      Project Rulison Due to the Release of Tritium   111
4-3   Radiation Dosages, in mrem, from the Hypothetical Use of
      Radioactive Natural Gas   114
4-4   Measured Peak Accelerations   116
4-5   Reported Damage Resulting from the Rulison Blast   116
6-1   Predicted Characteristics of the Rio Blanco Chimney   178
6-2   Comparison of Estimated Gas Productivity for Nuclear
      Stimulation Projects   179
6-3   Estimated Radioactivity Produced and Released in Project
      Rio Blanco Detonation and Flaring   181

FIGURE

2-1   Annual U.S. natural gas requirements, 1961-1990   24
2-2   Estimated cumulative U.S. natural gas requirements,
      1969-1990   25
2-3   Predicted sequential phases from detonation of nuclear
      explosive to minutes afterward   30

2-4	Effect of nuclear fracturing on well bore of reservoir model	31
3-1	Test site area	53
3-2	Project Gasbuggy predicted underground effects	58
4-1	Project Rulison area map	74
4-2	Schematic cross section of the Rulison area	104
4-3	Methane and carbon dioxide contents of Gasbuggy and Rulison gases	108
4-4	Methane and carbon dioxide contents	109
6-1	Project Rio Blanco experimental and development area location within the Piceance Creek-Yellow Creek drainage basin	171
6-2	Project Rio Blanco site map	172
6-3	Low-permeability natural gas areas of the Rocky Mountain states	175
6-4	Expected configuration of the chimneys and fracture zone	177
6-5	Oil shale areas in Colorado, Utah, and Wyoming	183
6-6	Schematic cross section of Piceance Creek Basin	184
6-7	Isopach map of 25-gallon-per-ton oil shale, and the Rio Blanco gas stimulation area, Piceance Creek Basin	185
6-8	Separation of nuclear chimney in Rio Blanco Project from overlying oil shale formations	187

Introduction

The purpose of this book is to analyze the effects of a new nuclear technology on the political, economic, and social life of a community and a state in the United States in terms which are intelligible to the average concerned citizen. To some, the technology appeared to hold a promise for relieving the threatening energy shortage; to others, it seemed to pose a hazard to the health and well-being of people and to their environment. This study, which was conducted by an interdisciplinary research team, is not intended to be a technical analysis or a scientific critique of the technology of nuclear gas stimulation, but it is rather an attempt to discern the broad implications of this technology in the context of the political, legal, and economic conditions in which it occurs.

We began this work as observers, with the intention of preparing an unbiased report on the facts of the case as they involved citizens, scientists, and government officials. However, during the course of our four-year effort, we found ourselves increasingly concerned about the potential hazards of large-scale nuclear gas stimulation and critical of the manner in which the technology was presented to the public. In particular, as a controversy over the Rio Blanco Project evolved in 1973, we found ourselves departing from the role of uninvolved chronicler and becoming, in fact, a part of the chronicle. As individuals concerned about possible long-term effects of this application of nuclear power, one of us decided to present testimony opposing the detonation of the Rio Blanco nuclear devices in Colorado.

We therefore present a point of view which is contrary to that of the former Atomic Energy Commission and others who are promoting the development of nuclear stimulation technology. The AEC and its industrial sponsors who were and still are involved in the program have a clear-cut interest to see this technology be used since they have spent a great deal of time, money, and effort on the endeavor and anticipate financial and other rewards if it is successfully implemented. We, on the other hand, as citizens and scholars who have studied nuclear gas-stimulation technology and its economic and ecological implications, have an equally clear commitment to point out what we believe are its shortcomings and potential dangers.

Boulder, Colorado   C. B. Wrenn
                                                    F. Kreith

Currently Professor of Chemical Engineering at the University of Colorado, Frank Kreith has been Research Engineer at the Jet Propulsion Laboratory at California Institute of Technology and a Guggenheim Fellow at Princeton.

Catherine B. Wrenn has taught in the Department of Political Science at the University of Colorado, where she is currently associated with the Department of Continuing Education.

Chapter 1

The Plowshare Program

We nuclear people have made a Faustian compact with society: we offer . . . an inexhaustible energy source . . . tainted with side effects that, if uncontrolled, could spell disaster.

Alvin Weinberg
Director, Oak Ridge Laboratory

The Plowshare Program was conceived in 1957. It was based on the premise that peaceful applications of nuclear energy could yield many benefits for human society. Yet today, 18 years later, experts still cannot agree whether nuclear explosives employed for peaceful ends may eventually result in damage of greater weight than the gains they provide. The potential danger of radioactive contamination of the environment by nuclear energy tests is still hotly debated among experts. Possible progress in such diverse fields as earth moving, mineral exploitation, energy production, and desalinization of sea water are being measured against the prospect that the planet may be made less habitable in the process. The political, administrative, and technological leaders of the Plowshare Program are forced to grapple with this dilemma. In the conflict between a vast nuclear potential and mounting environmental concerns, there are no ready answers.

As Plowshare developed, its emphasis was increasingly directed toward one specific industrial application: nuclear stimulation of natural gas. This study was therefore focused on Plowshare experiments aimed at the future production of natural gas through the use of nuclear explosives. The projects included were the Gasbuggy (New Mexico, 1967), Rulison (Colorado, 1969), and Rio Blanco (Colorado, 1973) detonations. In the chapters that follow, these projects receive detailed examination from technical, political, social, and economic points of view.

The story of the Plowshare Program is a continuing one. The final outcome of the Rio Blanco experiment is not yet known. Nor is it known how the functions of the dissolved AEC will be allocated within the new Energy and Resources Development Administration. This Administration, which may be the forerunner of a Cabinet Department of Energy and Resources, has combined under one roof the energy-research activities of the Atomic Energy Commission, the Department of the Interior, the Environmental Protection Agency, and the National Science Foundation. In addition, the reorganization calls for the establishment of a Nuclear Regulatory Commission to undertake the licensing and regulatory duties previously handled by the AEC. Thus, this study leaves its story incomplete. But an analysis of Plowshare projects and proposals undertaken by the AEC and its industrial cosponsors may foreshadow the direction that the entire Plowshare Program and other large-scale resource-development projects will take in the future.

LEGISLATIVE HISTORY: THE ATOMIC ENERGY COMMISSION

It was evident from the time that atomic devastation was unleashed on Japan at the close of World War II that there was no simple answer to the problem of control of nuclear energy. Congress had to devise legislation to control forces that had a direct bearing on human survival. There was no way to wish away the nuclear genie. Work had been done on nuclear development in Germany, in the Soviet Union, and elsewhere, and this could not be wished away. The United States had to attempt to maintain and control its monopoly, or at least continue its leadership, in the atomic energy field; this was an obvious military imperative. The United States offered to place atomic energy under international control through the Baruch Plan, introduced in the United Nations in 1946. Under the terms of the plan, however, no other nation would have been allowed either to begin or to continue atomic investigation except under strict and enforceable international control, as stipulated in the plan. Equally important civilian imperatives were to put to peaceful uses the technology underlying the manufacture of stockpiles of American atomic bombs and to establish strict controls against military destruction.[1]

The Atomic Energy Act of 1946 was the response of Congress. Although not in session on August 6, 1945, when President Truman announced the first use of an atomic bomb, Congress reconvened quickly. Perhaps because of this haste, the powers given to the AEC were unprecedented.[2] The newly established commission was given complete and exclusive control over ownership, production, and use of all atomic material, whether civilian or military. This control extended to all areas that could conceivably be linked to atomic development: research and development, programs on radiological health and safety, ore mining and refining, the granting of licenses, control over lands having radioactive resources on which atomic facilities were to be built, and even ownership of the facilities themselves. The Commisson was also responsible for the administration of any international agreements concerning atomic energy. All this was to be in addition to the AEC's control over bomb production and other military and security matters.

Aside from its monopoly in the atomic field, the AEC was given an extraordinary amount of operational independence. Although it seemingly was to work in rather close conjunction with the President and Congress, in time it became far less broadly accountable to them than were other federal agencies. By the terms of the 1946 Act, the President had the power to appoint Commission members with the advice and consent of the Senate; he could also remove members for cause before the expiration of their terms. The President had executive authority over the AEC budget, classified information, weapons quotas, and all disputes involving the Commission. In addition, his approval was required for all international agreements involving cooperation in atomic matters. The Act provided for Congressional supervision by establishing a Joint Congressional Committee on Atomic Energy to hold hearings and make recommendations on any proposed legislation relevant to atomic energy. The Act also

3

provided for the possibility of judicial review of any action undertaken by the Commission.

The Atomic Energy Act thus imposed restraints on the broad scope of the AEC. Since 1946, however, there have been powerful factors at work that have reduced the effectiveness of the original restrictions. Historical circumstances of the Cold War from 1946 on pushed the United States to maintain, in fact to strengthen, its leadership in military technology and materiel. The national effort in atomic weaponry was of paramount concern and received high priority. In the interests of national security, the AEC was given wide latitude in all phases of decision-making. Since the subject of atomic and nuclear energy is generally of such scientific complexity that no layman can readily understand the technicalities involved, a natural trend developed: to leave complex problems to the experts, the Atomic Energy Commission, and the President's scientific advisors. These advisors often had gained their competence in atomic matters under the AEC because of its monopoly in the field. Finally, the restraints written into the Act were weakened by the proliferation of bureaucracy within the AEC itself. As the Commission came to life, the addition of divisions, subdivisions, and advisory groups made it difficult to determine how decisions were made and by whom along the bureaucratic chain of command. Such difficulties persisted throughout the life of the AEC.

Most important of all, perhaps, was the fact that the Atomic Energy Commission was at once chief regulator and chief promoter of nuclear development. This situation may be considered a distinct violation of acceptable governmental organization, because the two functions may frequently conflict. If the AEC were to be interested primarily in advancing the cause of nuclear development, whether military or civilian, regulatory obligations would inevitably become secondary in importance and in the vigor with which they were pursued. Conversely, if regulatory action were the AEC's prime concern, proposed developmental activity would be thoroughly analyzed and cautiously undertaken. The legislative framework of nuclear development placed the AEC on the horns of a dilemma: how to satisfy the demand for exploitation of nuclear energy while meeting the cry for regulation to protect public health and safety. Perhaps the reorganization under the new Energy and Resources

4

Administration will help to resolve the problem, but this basic dichotomy plagued the AEC.

Recently, difficulties have been compounded by the need to comply with the National Environmental Policy Act of 1969. On the one hand, the AEC was under pressure to promote large-scale development of commercial nuclear reactors as a means of combatting the energy crisis. On the other hand, there has been pressure to halt such development until it is clear that no irreparable environmental damage will result. The AEC was forced to confront the problem directly as a result of the U.S. Supreme Court's Calvert Cliffs decision. The Court ruled that the AEC had failed to establish adequate guidelines for preparation of required environmental impact statements, and it demanded the Commission conform to the spirit and purposes of the National Environmental Policy Act.[3] The slowdown in the Plowshare Program is another indication of the fundamental dilemma.

LEGISLATIVE AUTHORIZATION OF THE PLOWSHARE PROGRAM

The strong environmental concerns of recent years had not yet developed when the enabling legislation for the Plowshare Program was passed. In 1954, the Atomic Energy Act of 1946 was amended to give statutory existence to the desire that the newfound source of energy "be put to peaceful uses for the benefit of all mankind." In the words of the 1954 Act:

> Atomic energy is capable of application for peaceful as well as military purposes. It is therefore declared to be the policy of the United States that
>
> a) the development, use and control of atomic energy shall be directed so as to make the maximum contribution to the general welfare, subject at all times to the paramount objective of making the maximum contribution to the common defense and security; and
>
> b) the development, use and control of atomic energy shall be directed so as to promote world peace, improve the general welfare, increase the standard of living and strengthen free competition in private enterprise.[4]

5

The section directly authorizing the Plowshare Program is in Chapter IV of the Act, under the title, "Research."[5] The AEC was authorized by the terms of this chapter to "exercise its powers in such a way as to insure the continued conduct of research and development and training activities . . . by private or public institutions or persons, and to assist in the ever-expanding fund of theoretical and practical knowledge . . . ." To accomplish these purposes, the Commission was empowered to make contracts, agreements, and loans for research and development activities related to (1) nuclear processes; to (2) the theory and production of nuclear energy; and to (3) the utilization of nuclear and radioactive material for medical, biological, agricultural, industrial, and commercial uses, including the demonstration of any advances in research on commercial and industrial applications. In addition, the AEC was made responsible for the promotion of safety and the protection of health during these research and production activities.

The 1954 legislation authorized the AEC to make grants for the construction and operation of reactors at universities and hospitals for educational and training purposes. When it found that private facilities were inadequate, it could use its own facilities to conduct the desired research so long as such action appeared appropriate to the overall goal of developing atomic energy for peaceful uses. In any agreements made, the Commission was, again, responsible for including provisions to protect health and property, and for requiring such reporting and inspection as was necessary to fulfill this responsibility.

On May 5, 1955, the AEC made use of its new powers and announced the start of two new nuclear programs. The first, the Power Reactor Demonstration Program, provided substantial federal assistance to utility companies to stimulate the building of nuclear reactors. The administrative and legal basis for the commercial generation of power by nuclear means was thus established. The second program was Plowshare, envisioning the use of nuclear explosives in industrial applications. Of the two programs, Plowshare was incorporated into the organizational structure of the AEC far less rapidly. This was at least in part due to the fact that the technology required for the industrial use of nuclear explosives had not kept pace with the technology involved in the use of nuclear reactors for the commercial generation of power. To this day, the Plowshare Program,

6

combining the resources of the AEC and private industry, has not fulfilled the potential envisioned for it by Congress.

THE PLOWSHARE PROGRAM: ORIGINS AND ORGANIZATION

The first informal discussions of a program designed to meet Plowshare purposes were held in November 1956 at an in-house conference of the Lawrence Radiation Laboratory in California. The following year, the first Plowshare Symposium was held—its subject was "Industrial Uses of Nuclear Explosives." According to Richard Hamburger, Assistant Director of the Office of Peaceful Nuclear Explosives (OPNE) of the Atomic Energy Commission, the basic concepts that have since guided Plowshare emerged at the in-house conference.[6] In June 1957, the AEC established a new research and development program and created the Division of Peaceful Nuclear Explosives to direct and administer it. This division, under the leadership of James Schlesinger, became the Division of Applied Technology. When necessary, that group could draw upon other branches of the AEC, including the Divisions of Biology and Medicine, of Operational Safety, of Public Information, of Military Applications, of International Affairs, and other headquarters groups.

The operational division of the AEC that has been responsible for conducting all nuclear explosions, including Plowshare experiments, is the Office of Peaceful Nuclear Explosives at Nevada, known as NVOO. In the last few years, NVOO has been responsible also for assisting private corporations in working out experiments in tests for various industrial applications of nuclear explosives. This is an instance of the AEC's dual role—its legal authority both to promote and to regulate atomic and nuclear matters on an everyday, operational level. In carrying out these two functions, NVOO worked with the U.S. Public Health Service, which is also charged with safeguarding the public against any adverse effects resulting from use of nuclear explosives. The task of informing the public about the radiological precautions undertaken in Plowshare Program experiments has often also fallen to the Public Health Service. Assistance in safety operations was provided by the

Environmental Sciences Service Administration (ESSA), the U.S. Bureau of Mines of the Department of the Interior, the U.S. Geological Survey, and the U.S. Coast and Geodetic Survey.

The major technical efforts since 1957 have been made at the Lawrence Radiation Laboratory, but significant work has been carried on at other national laboratories as well. Additional government agencies involved in these efforts include the Bureau of Mines, ESSA, the U.S. Geological Survey, and the U.S. Coast and Geodetic Survey.

The principles underlying the effort to achieve industrial development were to be found first in the link between the AEC and private industry and secondly in industry's willingness to stake its resources on future commercial development and sale of nuclear-generated products.

THE SEARCH FOR DIRECTION

Early Plowshare discussions projected glowing possibilities in the development of nuclear energy, and generally took an optimistic view of its potential liabilities. At the Second Plowshare Symposium (San Francisco, 1959), Edward Teller suggested two main activities to be pursued in the peaceful uses of nuclear explosives. These were mining and building. He defined them so broadly, however, that the implication was that nuclear energy could be used in almost any large building or excavating project. A second section of Teller's statement concerned three major drawbacks that he foresaw: radiation hazard, seismic effects, and the size of explosive required.[7]

Throughout the early years of experimentation under Plowshare, and indirectly in coordination of experiments with the Department of Defense, the AEC's attitude was that the use of nuclear explosives would be considered only in terms of proven technology, safety factors, and economics. Aspects outside the scope of a technical program—political, sociological, and psychological considerations—were not matters of AEC concern. Given this assessment of its role, and the fact that radiation exposure was not looked upon as an exceptionally unsafe or limiting consideration, it is not surprising that both the AEC and its scientists, committed to the expanding use of nuclear

8

power, projected plans for literally earth-shaking development.

More than 150 nuclear explosions—atmospheric, surface, and underground—were undertaken prior to the time of any test specifically designed for the Plowshare Program. These detonations were carried out primarily for weapons-development purposes, but they also provided a store of information which Plowshare scientists could use in evaluating the impact of nuclear explosions in nonmilitary applications such as rock fracturing, cavity formation, heat transfer to surrounding materials, and containment of radioactivity.[8]

Between 1958 and 1961, the AEC conducted a number of experiments in crater research with conventional explosives because during this time the United States was restrained from nuclear testing by the International Nuclear Test moratorium, which began in late 1958. In December 1961, following the end of the moratorium, Project Gnome, the first specific Plowshare experiment, was carried out. Its primary scientific objectives were to make isotope recovery studies, to determine heat-recovery possibilities, to perform a neutron physics experiment, to record information on shock pressures on mineral rock and organic material samples, and to conduct seismic studies to provide information on the structure and properties of the Earth's crust and mantle east of the Rocky Mountains. Project Gnome involved the detonation of a nuclear explosive with a yield of 3.1 kilotons about 25 miles southeast of Carlsbad, New Mexico. The nuclear device was situated in a salt formation 1,200 feet below the Earth's surface. After detonation, a cavity of about 960,000 cubic feet was produced, with some 2,400 tons of rock melted, and "most of the non-gaseous radioactive residue . . . trapped in the mixture of rubble and once molten salt below the chamber."[9]

From 1962 onward, several specific Plowshare experiments were carried out to test the effects of various kiloton devices exploded at various depths in different media. In each test, the general purposes were to develop techniques for nuclear excavation, obtain data on cratering effects, and collect information which would assist in prediction of safety.[10]

One of the first industrial applications proposed to the AEC came from the Santa Fe Railroad and the State of California. The proposed plan, named "Carryall," was to determine the feasibility

9

of using nuclear explosives to excavate a transit route through the Bristol Mountains in the Mojave range, wide enough to accommodate a railroad right of way and a highway access route. This project did not go beyond a feasibility study because of the incompatibility of the technology with the highway construction schedule. Another plan given early consideration was Project Travois, which would have used nuclear explosives to develop a quarry-dam system for the Twin Springs area near Boise, Idaho, and would have been carried out under the joint auspices of the AEC and the U.S. Army Corps of Engineers. The AEC cited a shift in federal priorities as the reason for inaction in this case.[11]

In 1965, the AEC began an environmental study under Plowshare aegis to determine the feasibility of a man-made, nuclear-excavated harbor at Cape Thompson, Alaska. At the same time, a similar study was requested by the Government of Australia for the Cape Kerauden area. There was a lack of action in both of these cases because it was feared there would be severe ecological damage as well as radiation hazards.

Another project, proposed in 1967 by the Kennecott Corporation, was to be concerned with *in situ* leaching of low-grade ore. Kennecott was prepared to fund half of the project if the AEC would fund the other half. This proposal, called "Project Sloop," envisioned the creation of chimneys or subterranean voids through the use of nuclear explosives. The chimneys were expected to be filled with crushed rock as a result of the blasts, and acid would have been used to percolate through the crushed rock. The resulting liquid, containing dissolved ore, was to be pumped to the surface, where the ore would then have been captured. This project failed to materialize when the AEC was unable to fund its half of the proposal. Budget cuts in the funding of the Plowshare Program were cited as the reason for inaction in this case.[12]

Another feasibility study, to have been completed by July 1, 1970, was on the possibilities of nuclear management and conservation of Arizona's water resources. A system of canals, reservoirs, and dams was proposed, to be engineered by nuclear methods and jointly sponsored by the AEC, the Interior Department, and the State of Arizona. After the United States' decision to adhere to the International Test Ban Treaty prohibiting the explosion of nuclear devices if radioactive contamination in the atmosphere would cross national boundaries, this project was abandoned.

It is possible that the AEC failed to act on these and other proposed Plowshare projects as much because of a lack of specific knowledge about nuclear technology in industrial applications as because of shifting federal budgetary priorities, or other reasons that were cited for inaction or postponement. In any case, the inability of the AEC to generate Plowshare momentum was a continuing source of concern to those in the Division of Peaceful Nuclear Explosives as well as to some members of the Joint Congressional Committee on Atomic Energy. Rep. Craig Hosmer (R., Calif.), who was a member of the Joint Committee, pointed out that the Plowshare Program was dragging compared to power-reactor development. Reasons for inaction from his point of view included the fact that industry was not being brought into the picture initially and that there was firm opposition within the Executive Branch to the Plowshare concept. Specifically singled out as opposing the Program were such diverse agencies as the Bureau of the Budget, the State Department, and the Arms Control and Disarmament Agency. Representative Hosmer placed the blame for inaction on these agencies which "share a paranoic distrust and abhorrence for Plowshare which they cannot divorce in their minds from the weapons program," and on the "so-called liberal who views any nuclear application in terms of a mushroom cloud."[13]

To help overcome such opposition Hosmer repeatedly introduced legislation to amend the Atomic Energy Acts to make Plowshare services available on a commercial basis. Such legislation, if passed, would have authorized and ordered the AEC to go beyond its legislative authorization, which permitted research and development only. This kind of proposed legislation repeatedly failed to pass, even with the bipartisan support of all House members on the Joint Committee. The effort to recognize Plowshare as a business venture with defined and standardized nuclear services that the government could render, with established price lists and provision for regulatory control, now depends on the directives of the Energy Research and Development Administration and the willingness of Congress to provide such enabling legislation as may be required.

FUNDING OF THE PLOWSHARE PROGRAM

Perhaps a combination of the difficulties that Representative Hosmer explained led to the great gaps between Plowshare Program cost projections for the fiscal years 1967 through 1971 and the amounts that were actually appropriated. The AEC, in providing the projected cost figures, was careful to point out that the amounts given represented its own internal estimates and plans, and did not signify Executive Branch policy approval. Their point was well taken. In April 1965, the AEC's funding for the Plowshare Program in 1967-71 was projected as $204,700,000. The estimate was given in response to questions from industrial representatives who wanted assurance of a relatively long AEC commitment to Plowshare. The AEC at that time was able to provide an overall cost estimate but was unable to give specific cost breakdowns.[14]

The AEC's projections in 1965 were clearly somewhat optimistic. The road from Commission request to the Executive Branch for budget formulation, then to the House Appropriations Committee, and then to the actual final appropriation appears to have been long and hard. The funding process obviously assumed directions which neither the AEC nor the Joint Congressional Committee on Atomic Energy was able to foresee or control. No doubt a partial explanation of the gap between expectation and reality is that this was a period of general economic recession in the United States, but in part it can also be attributed to a growing sense of uneasiness about the liabilities of using nuclear power, even for peaceful purposes. In 1967, James Terrill of the Consumer Protection and Environmental Health Service voiced his three main concerns about the Plowshare Program at the Public Health Symposium as: (1) radioactivity connected with the production of natural gas; (2) radioactive contamination of consumer products, such as strontium-90 in milk; and (3) radionuclides in our biosphere, including increases in concentrations of tritium, carbon-14, and krypton-85.[15] Such liabilities, pushed to the background by early Plowshare enthusiasts, were now being considered by people both inside and outside of the government. The budget cuts and lower final appropriations for the Plowshare Program may reflect in part this growing public concern.

12

THE EMPHASIS ON NUCLEAR STIMULATION OF NATURAL GAS

Internal and external pressures apparently combined to direct the major part of Plowshare resources toward achieving nuclear stimulation of natural gas on a commercially and environmentally viable basis. The efforts of the Joint Committee to the contrary, a combination of budgetary restrictions, concern for radiological safety and environmental protection, economic feasibility, the Test Ban Treaty, and the general question of national priorities had all slowed the Plowshare Program to a snail's pace. Though there was considerable work done on the feasibility of fracturing oil-shale rock by nuclear detonations and extraction of the oil through on-site retorting, there has been little current emphasis on this kind of technology. In the mid-1960s, the AEC, Department of the Interior, CER-Geonuclear Corporation, and the Lawrence Radiation Laboratory did prepare a comprehensive study of the Piceance Creek Basin near Meeker, Colorado, relating to recovery of oil from shale. The resulting proposal, called "Bronco," is mainly of interest now because there is a continuing contest between representatives of oil shale and natural gas interests who wish to develop potential resources in the same geographical area.[16] The oil shale interests, however, are apparently not now seriously considering nuclear means of extraction.

It appears that the only project emerging in the past six years from the Plowshare Program with anything remotely resembling healthy vigor is that of nuclear stimulation of natural gas. Project Gasbuggy, carried out near Farmington, New Mexico, in 1967, was the AEC's first completed joint venture with industry to stimulate natural gas. In 1963, the El Paso Natural Gas Company, the Interior Department, and the AEC jointly studied with the Lawrence Radiation Laboratory the feasibility of using a nuclear device to stimulate a tight natural gas reservoir. The results, published in 1965, were that such an experiment could be conducted safely and that the experiment was necessary to further technical understanding.

Project Gasbuggy took place on December 10, 1967. A 29-kiloton nuclear device was detonated 4,240 feet deep in the San Juan basin in northwestern New Mexico. A fusion bomb was used, creating a chimney of broken rock about 333 feet high and with a volume of about 2 million cubic feet, 3,900 feet below the surface.[17] According to testimony given by Sam Smith, Assistant Vice-President of El Paso, on March 23, 1971, before the Joint Committee on Atomic Energy, Gasbuggy successfully demonstrated that the use of nuclear explosives resulted in a substantial increase in both rate and quantity of natural gas production, that the nuclear chimney could be reentered through the casing used to house the device, and that the size of the chimney was accurately predicted. He also testified that the chemical and radioactive content of the gas had been measured so that accurate predictions about gas quality in future tests could be made. Smith further stated that a study by the Oak Ridge Laboratory and the El Paso Natural Gas Co. demonstrated that if a large amount of nuclear stimulated gas was used commercially, the average radiation exposure of a California user would be less than one millirem per year. Smith did concede the experiment demonstrated the need for a nuclear explosive which produced minimum residual tritium, or radioactive hydrogen. The fusion bomb used in Project Gasbuggy had generated about four grams of tritium. Tritium, which is easily absorbed biologically, is a long half-life isotope (12.3 years).[18] In production tests, some of the trapped gas in the Gasbuggy chimney was flared, or burned off, a number of weeks later. The AEC, El Paso Natural Gas, and the U.S. Bureau of Mines estimated that over a period of 20 years, the cumulative gas production from the Gasbuggy field would be about 1 billion cubic feet. This is about 20 percent of the total gas under the 160 acre site. By comparison, conventionally drilled wells from a nearby site were estimated to be capable of producing about 125 million cubic feet over a 20-year period. This is an eightfold increase in production.[19]

The AEC has stated that radioactive concentrations, especially tritium, were less than expected, but emphasized that a new explosive design might be necessary in general to minimize the production of tritium.[20] Public Health Service studies showed a tenfold increase in radiation downwind of the test site. The AEC did not mention the Public Health Service finding in providing information to the public on Project Gasbuggy but did state that monitoring for radioactivity in feeder lines from nearby wells showed no increased radioactivity attributable to Gasbuggy.[21]

The next industrial-AEC venture in the nuclear stimulation of natural gas, Project Rulison, involved a fission rather than a fusion bomb in order to reduce as much as possible the concentration of tritium. The Rulison Project was undertaken by the AEC, Austral Oil Company, and the U.S. Department of the Interior, with CER-Geonuclear Corporation of Las Vegas acting as program manager. The Los Alamos Scientific Laboratory provided technical direction for this experiment.

A 40-kiloton device was detonated on September 10, 1969, in Garfield County, near Rifle, Colorado, at a depth of 8,430 feet. The general objectives of the Rulison blast were to gather additional engineering data on nuclear stimulation of natural gas and to determine changes in gas production and recovery rates stemming from the larger kiloton device employed in this project. Additional objectives were to evaluate various techniques for reduction of radioactive contamination of the gas, and to generally supplement the Gasbuggy data and thereby provide a basis for predicting results for future experiments.[22]

Post-detonation testing for the Rulison, as for the Gasbuggy experiment, involved flaring of the gas after the period of time considered adequate to allow radioactive isotopes to decay to safe levels. The flaring process burns off radioactivity by producing water vapor, which is then released into the atmosphere. In October, flaring was begun, and by the end of 1970, the Atomic Energy Commission announced that the quality of the gas was improving compared to that obtained in October. The reduced radioactivity was attributed to fresh gas flowing into the well from the surrounding fractured rock, which diluted the original concentrations. In its 1970 report to Congress, the AEC stated that the tritium content of the gas was lower than expected, and considerably less than that found in the Gasbuggy gas. However, the amounts of cesium-137 and strontium-90 generated were large in comparison to Gasbuggy levels. Both cesium-137 and strontium-90 can be absorbed biologically and perpetuated in the food chain through vegetation to livestock and eventually to humans.[23]

In summary, the AEC concluded after the Rulison experiment that tests showed no radioactivity above background levels in streams and drinking water, and that there were sufficient techniques available to insure that production testing could proceed safely. There are some scientists, however, who believe that *any* radioactivity in this type of gas could be dangerous because, since it can be concentrated in the food chain, the dilution of radioactivity in nuclear-stimulated natural gas, if commercially used, would merely spread a lower dose of pollution to an increased number of people.[24]

While Project Gasbuggy took place with a minor degree of public concern, the Rulison blast was the focus of public apprehension. Legal attempts were made on the basis of both environmental and safety grounds to prevent both the initial detonation and the later flaring of the gas. Requests for court orders to halt the Rulison explosion and flaring were denied. The grounds given were that the AEC was reasonably exercising its legal authority to carry out nuclear research and development, and that the protesting groups had failed to show that the provisions made for protection of public health and safety were inadequate. [25] Efforts to secure a reversal of the decision by appeals to higher courts were also unsuccessful. At most, the legal suits served to alert the AEC and its industrial partners that there were public interest groups who were not resigned to quietly accept all proposed nuclear experiments in Colorado.

The next Plowshare Program experiment, Miniata, was scheduled for early June 1971 at the Nevada Test Site. Miniata was delayed twice because of difficulties with the 17,500-foot shaft in which the device was to be detonated. The shot was actually fired in September 1971, producing an explosion with the force of 80,000 tons of TNT. This is approximately four times the force of the bomb dropped on Hiroshima. The Miniata nuclear device was a small one, 9 inches wide and 30 feet long. It was designed to slide down standard drill casings that are used in natural gas fields and to withstand pressures at the 10,000-foot level, where it would be used in natural gas recovery projects. The AEC termed the Miniata shot "successful," both technically and in that no radiation was released into the atmosphere. Shock waves, gently felt by observers 12 miles away, were within the predicted range.[26]

Meanwhile, there are a number of proposed projects which would indicate that the AEC had settled on nuclear stimulation of natural gas as the area for proving that the Plowshare Program was a viable undertaking. The proposed Wagon Wheel Project has been under serious consideration, though its present status remains very much in doubt due to highly organized and vocal local opposition and questionable future Congressional funding.

As proposed, Wagon Wheel would take place in the Pinedale Unit of Sublette County in southwestern Wyoming. El Paso Natural Gas Company, in conjunction with the Lawrence Radiation Laboratory, prepared a plan for the sequential firing of a series of blasts, from bottom up, in a single well. Predictions indicated that sequential firing would triple the gas production that results from the firing of a single nuclear blast, and would significantly reduce tritium concentrations in the gas and associated water vapor that were found in the earlier Gasbuggy test. These predictions, however, were made prior to the Rio Blanco experiment, which apparently left serious questions as to the technical feasibility of sequential firing.

The most recent experiment under Plowshare, presumably a crucial developmental step prior to implementation of Wagon Wheel, was Rio Blanco. This project, with scheduled dates for detonation varying from late 1971 to late 1973, was finally carried out in May 1973. The AEC's industrial associates for this venture are the Equity Oil Company of Salt Lake City and CER-Geonuclear of Las Vegas. The partners pooled their efforts to set off three nuclear devices in the same well shaft at a site in the Piceance Creek Basin near Meeker in western Colorado. The projected development of the whole Rio Blanco field includes plans for more than one hundred blasts to be carried out over a number of years. The AEC and its industrial partners accepted the fact that the public must be informed of, if not wooed into acceptance of, nuclear projects, and consequently they held a number of public hearings on the proposal. The AEC specifically held hearings in Meeker, Colorado, and in Denver in March of 1972 to allow expression of opinion, which turned out to be both favorable to and highly critical of the experiment. Although some who testified at the hearings maintained that the deck was stacked in favor of the sponsoring agencies and industry, this kind of public hearing was unique in AEC history.[27]

The hearings reflected, at least to some extent, the thinking of the then new Commissioner of the AEC, James Schlesinger.[28] Schlesinger made it clear that the Atomic Energy Commission would fulfill its responsibilities to the nuclear industry in a somewhat different fashion than had been the case in the past. In a speech on October 20, 1971, he indicated that the Commission would provide technical options and see to it that nuclear technology was safely and appropriately used. He further said that the nuclear industry must learn to fight its own political, social, economic, and commercial battles, and must work toward the development of a comprehensive set of safety criteria and industrial standards. The AEC hearings in Denver and Meeker on Project Rio Blanco were at least a partial step in this direction.

During the course of the Rio Blanco hearings, commercial partners of the AEC discovered that there are indeed political, social, and commercial battles to be fought. The Governor of Colorado at that time was John Love; he appointed the Governor's Commision on Rio Blanco to obtain advice on the merits or disadvantages of proceeding with Rio Blanco. A Colorado legislator introduced a bill in the State House (H.B. 1018) designed to create a Colorado Atomic Energy Act, in an unsuccessful attempt to give Colorado control over its own nuclear future. Such proposed legislation had virtually no chance of passage, as it would have directly clashed with the AEC contract with the State of Colorado, but it did indicate a desire in the state to exercise some control over Colorado's environmental destiny.

Not the least of the opposition to Project Rio Blanco expressed at the hearings came from The Oil Shale Company (TOSCO), which vigorously opposed any nuclear stimulation of natural gas in the Piceance Creek Basin at this time. Their opposition was based on the alleged incompatability of the simultaneous conventional recovery of oil shale and nuclear stimulation of natural gas. In their view, the potential of oil shale as an energy source in the area is far more valuable than that of natural gas.

Although evaluation of the results of the Rio Blanco project have not been completed, at this time available results are not very encouraging for the economic and technical viability of underground nuclear stimulation of natural gas. Moreover, since former President Nixon proclaimed in 1973 "energy self-sufficency for the United States by the 1980's" as a national goal, the opposition of the burgeoning oil-shale industry may become increasingly important in planning for the development of energy sources. These developments, added to the opposition of the citizens of Colorado, may well halt further nuclear stimulation of natural gas in the Rocky Mountain West. In short, any future plans of the AEC and its industrial sponsors may be overruled by political, economic, social strategic considerations over which the proponents of nuclear stimulation have little control.

NOTES

[1]See Richard Curtis and Elizabeth Hogan, *Perils of the Peaceful Atom*, 1970, for an analysis of this point of view. The following discussion also relies heavily on their work in covering the legislative history of the AEC.

[2]The Atomic Energy Act of 1946 (Public Law 585, 79th Cong., 60 Stat. 755-75, 42 U.S.C. 1801-19). For the legislative history index to Public Law 585, see *Atomic Energy Legislation Through the 91st Congress, 2nd Session.* Joint Committee on Atomic Energy, Congress of the United States, Jan. 1971, Appendix C, p. 272.

[3]For analysis of the *Calvert Cliffs* decision and its impact on AEC procedures, see pp. 132-133.

[4]*Atomic Energy Act of 1954*, 42 U.S.C. sec. 2011.

[5]*Ibid.*, Chapter 4.

[6]Speech at Germantown, Maryland, July 11, 1969; in SWRHL-82, 1970.

[7]Edward Teller, Univ. California Radiation Laboratory (UCRL-4659).

[8]Carl Gerberger, Richard Hamburger, E. W. Seabrook, "Plowshare," U.S. Atomic Energy Commission (USAEC), Division of Technical Information, 1966 (pamphlet), p. 24.

[9]*Ibid.*, p. 26.

[10]*Ibid.*, pp. 26, 32, 34; and USAEC, *The Plowshare Program during 1968*, pp. 3-4.

[11]See John Tonan, Lawrence Radiation Laboratory, Plowshare Symposium, (CONF 690609), Jan. 14, 1970.

[12]See S.D. Michaelson, Kennecott Copper Corp., Salt Lake City, Utah, Plowshare Symposium (CONF 690609; CONF 700101, vol. I), Jan. 14, 1970.

[13]Symposium on Public Health, Southwestern Radiation Health Laboratory (SWRHL-82), 1969.

[14]See Hearings before the Joint Committee on Atomic Energy, "Peaceful Applications of Nuclear Explosives—Plowshare," 89th Cong., 1st Session, Jan. 5, 1965, pp. 509-512, and the *Budget of the United States, Appendices Fiscal Year 1967, 1968, 1969, 1970, 1971, 1972.*

[15]Plowshare Symposium (CONF 680330), 1967.

[16]See *The Bronco Oil Shale Study*, Oct. 13, 1967. Springfield, Va.: Clearing House for Federal Scientific and Technical Information (PNE-1400).

[17]USAEC, *Annual Report to Congress, 1970.* Jan. 1971, p. 197.

[18]Milo D. Nordyke, Plowshare Symposium (CONF 690609), Jan. 14, 1970.

[19]USAEC, *The Plowshare Program during 1969*, p. 4.

[20]USAEC, *The Plowshare Program during 1968*, p. 6.

[21]USAEC, *Annual Report to Congress, 1970*, p. 197. For a highly critical report of the Gasbuggy operation, see H. Peter Metzger, "Project Gasbuggy and Catch-85," *New York Times Magazine*, Feb. 22, 1970.

[22]USAEC, *The Plowshare Program during 1969*, p. 4.

[23]See Nordyke, "Technical Status Summary," UCRL 72332.

[24]See Cheryl Haynes, "About Rio Blanco; Three Questions for the AEC," *dear earth*, June 1971, pp. 17-19.

[25]The legal proceedings prior to and after the initial Rulison shot are discussed below; see pp. 91-101.

[26]Press release by A. Dean Thornbrough of the AEC Office of Peaceful Nuclear Explosives, Las Vegas, Nevada, Sept. 8, 1971.

[27]For details of the Meeker and Denver hearings, see pp. 135-140.

[28]Schlesinger, who was something of a bueaucratic troubleshooter during the Nixon Administration, was subsequently named Secretary of Defense. Dr. Dixie Lee Ray was given the post of AEC Commissioner, a job that she held up to the time of reorganization and the creation of the Energy Research and Development Administration.

Chapter
2

Natural Gas: The Nuclear Route from Reservoir to Pipeline

Though there are differences of opinion about the wisdom of continuing the Plowshare Program, there is little argument about the current energy crisis. Unless new energy sources are found and made available for general use, Americans will be forced to curb their demand for energy. One way of partially supplying the demands for fuel and chemical products may be through nuclear stimulation of natural gas resources. There are many problems to be resolved before gas released by nuclear technology can be sent on its way through the pipeline to users. A survey of these problems on the following pages will serve to sketch the broad range of considerations which are integral to the nuclear development of natural gas resources.

THE NATIONAL NEED

In some parts of the United States, suppliers of natural gas have been unable to meet the demand and have been forced to limit sales to existing customers and to refuse to accept new ones. In early 1973, Denver public schools were temporarily forced to go on a three-day school week because of insufficient gas supplies to heat school buildings. On a nationwide scale, it has become

21

increasingly apparent that natural gas has not been available in sufficient quantity to meet the demands of power plants, which often have been faced with short supplies. Since natural gas combustion produces less air pollution per unit power output than other fuels and is cheaper per heating unit than any except coal, it is particularly attractive to industries such as electrical generating stations which have serious emission-control problems. According to John F. O'Leary, the AEC's Director of Licensing, natural gas production was increasing constantly until 1969 and appeared to be one of the solutions for fulfilling energy requirements. In 1969, however, the supply was "virtually shut off."[1] From the natural gas industry's perspective, this situation was caused by the Federal Power Commission, which set unrealistically low gas rates and prices that in turn produced insufficient revenue for exploration and development of new resources.

Whether the current shortage is simply the end result of the desirability of natural gas as a cheap and clean source of heating and energy, the result of poor governmental or industrial planning, or a combination of these factors, predicted future needs present an awesome picture when measured against existing supplies in the United States. The Future Requirements Committee of the American Gas Association has assembled data on future needs for natural gas in the United States. The Mineral Resources Institute of the Colorado School of Mines has made a comprehensive assessment of supplies. Other agencies and individuals have compiled similar surveys. All of them predict serious supply and demand differentials in the immediate future.[2]

All of these predictions are based solely on the supplies and needs of the United States. If worldwide natural gas needs and resources were considered, the outlook might be better. But foreign supplies must be weighed in the light of national and international politics and economic situations, various national defense programs, and differing cost factors. All of these factors make worldwide predictions subject to a great degree of uncertainty.

Gas-use predictions within the United States are based on a number of assumptions. These may or may not hold true over a period of several years, and they have their own degrees of uncertainty. However, it is generally assumed that: (1) the current relationships of natural gas costs to those of competing fuels will remain the same; (2) there will be adequate supplies of gas available for predicted needs; (3) the percentages of natural gas required by different industries and individual segments of the market will not undergo marked changes from present figures; and (4) an annual average national economic growth rate of 3.6 percent will occur over the time period for which predictions are made. The assumption implied here is that there is a direct relationship between energy demand and the Gross National Product which reflects national economic growth.

These assumptions hypothesize that the life styles, priorities, and material appetites of Americans will continue to expand as they have in the past, or at least will remain essentially unchanged over the 30-year projection period. Should any of these assumptions prove to be inaccurate, there would necessarily have to be revisions made in resource supply and demand estimates. There is, for example, some reason to believe that economic growth over the next few years will not reach 3.6 percent. If this is in fact the case, a thorough-going reassessment of the stated assumptions would be required.

Underlying the above assumptions is the general American conviction that growth automatically means progress, and necessarily results in better lives for us all. During recent years, however, this conviction has not been uniformly shared by all. The influence of ecology, with its emphasis on delicate systems in balance, and the growing awareness of the limited resources of our "spaceship earth" are some of the factors which may inhibit the growth of unrestrained material consumption. It is not beyond the realm of possibility that Americans within the next twenty years will revise present patterns of consumption either by necessity or by conscious effort, or both. Such a change in life style could well involve a declining birth rate and a consequent lessening demand for goods, services, and the raw materials required to produce them. In spite of such considerations, however, predictions are traditionally most safely made when based on already known empirical data. On this basis, all predictions point to the need for more natural gas supplies in the future.

Figure 2-1 shows the actual gas requirements for the United States from 1961-1971 and the estimated requirements for 1975-

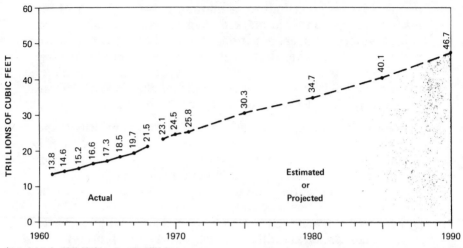

*Actual data for 1961-1963 based on the 1966 Survey.

Figure 2-1. Annual U.S. natural gas requirements, 1961-1990. ("Future Natural Gas Requirements of the United States," Vol. 3, Sept. 1969, p. 8, from the Future Requirements Committee of the Gas Industry Committee, published by the Future Requirements Agency, Denver Research Institute, Univ. Denver.)

1990. The same data are used in Figure 2-2 but are presented in terms of cumulative demand. It is estimated that the total cumulative requirement from 1969 to 1990 will be 760.9 trillion cubic feet of natural gas. This demand is then broken down into uses by market segments in Table 2-1.

The Federal Power Commission has been asked to consider limiting or even prohibiting industrial uses of natural gas. The reduction or exclusion of industrial demand from the percentages in Table 2-1 would significantly reduce the total demand for natural gas.

The Colorado School of Mines' Mineral Resources Institute has analyzed the other side of the picture, the supply of natural gas. The Potential Gas Committee that drafted the report was made up of 150 industrial and government figures who studied potential supplies of natural gas in the United States exclusive of

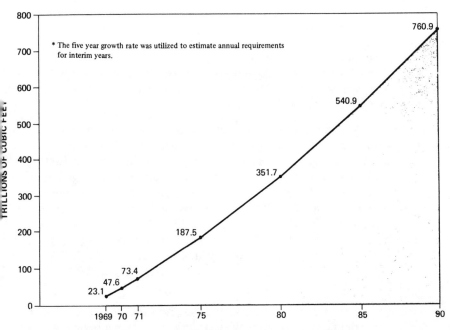

Figure 2-2. Estimated cumulative U.S. natural gas requirements, 1969-1990. ("Future Natural Gas Requirements of the United States," Vol. 3, Sept. 1969, p. 7, from the Future Requirements Committee of the Gas Industry Committee, published by the Future Requirements Agency, Denver Research Institute, Univ. of Denver).

TABLE 2-1
UNITED STATES NATURAL GAS REQUIREMENTS BY CLASS OF SERVICE
(1968, 1980, 1990)

	Year 1968		Year 1980		Year 1990	
Class	Volume	Percent of total	Volume	Percent of total	Volume	Percent of total
Residential	4.7	21.9	7.0	20.2	9.2	19.7
Commercial	1.7	7.9	3.0	8.7	4.5	9.7
Industrial	6.4	29.8	11.8	34.0	17.4	37.2
Interruptible	4.5	20.9	7.3	21.0	9.7	20.8
Other	1.4	6.5	2.3	6.6	3.0	6.4
Field Use	2.8	13.0	3.3	9.5	2.9	6.2
TOTAL	21.5	100.0%	34.7	100.0%	46.7	100.0%

Source: "Future Natural Gas Requirements of the United States," Vol. 3, Sept. 1969, p. 8, from the Future Requirements Committee of the Gas Industry Committee, published by the Future Requirements Agency, Denver Research Institute.

Hawaii. Their figures are not reassuring when compared to the predicted demand over the next twenty years. The 1970 report estimated a total potential supply of 1,178 trillion cubic feet (tcf) in the United States. This is broken down into "probable supply" (gas associated with existing fields), "possible supply" (gas in yet undiscovered fields in areas of already established production), and a "speculative" category which concerns "gas supply in new territories where there is no present production and the estimates are based on a minimum of information."[3]

The authors of the report point out that more than 60 percent of the potential gas resources are either more than 15,000 feet below the surface or under the oceans bordering the continental United States or in Alaska. Though the Potential Gas Agency Committee felt that this natural gas could be used commercially when found, the difficulties and expense involved in finding it would clearly be great. The estimated 1,178 tcf is approximately four times the 290 tcf of known natural gas reserves. For comparison, in 1969, the United States produced a record 22 trillion cubic feet of natural gas, while estimated demand for 1975 is 30.3 trillion cubic feet.

Although these figures on potential reserves indicate that the cumulative demand of 760.9 trillion standard cubic feet projected for 1990 by the American Gas Association could be met, it would create serious problems. Natural gas producers have insistently claimed that prices set for natural gas are too low to permit or encourage the exploratory activities which are required to locate new supplies. A great deal of potential existing gas supply is now so inaccessible that retrieval would require costly and sophisticated technological methods and equipment. Such factors help to account for the virtual "shut-off" mentioned by O'Leary. Though the Federal Power Commission has allowed increases in the price of gas, from the producers' point of view, these steps are too little and perhaps too late.

In addition to lowering the national demand for resources, other alternatives are also receiving consideration. There is the possibility that "synthetic" natural gas production could add to supplies. "Synthetic" natural gas means production of a chemically equivalent product, which is equally usable for power and heating. Coal gasification is the most likely synthetic process, but there are a number of factors which have slowed the effort. Technological uncertainty and economic feasibility are clear impediments. There is, in addition, concern about ecological, labor, and safety factors involved in the coal-mining process. In addition to these considerations, environmental pollution problems are associated with the transportation and distribution of the synthetic gas itself. The overriding difficulty, however, in all synthetic processes under consideration, is the lack of an available technology to produce economically competitive products.

The level in the price of gas set by the Federal Power Commission, prior to widespread concern over the energy crisis, was somewhere between 20 to 30¢ per thousand cubic feet at the wellhead. Transportation costs in some cases added as much as 30¢ per thousand cubic feet to the delivered product. Recently, the Federal Power Commission has attempted to let the price of gas gradually seek its own level on the basis of supply and demand and production costs.[4] Whenever the Commission has allowed natural gas producers to sell gas at unregulated prices, this has induced rapidly rising prices paid by the consumer for the delivered product.[5] If these high prices are maintained or increased, alternative synthetic resources may well become economically competitive. However, the effect of pervasive inflation at all levels of the economy makes all predictions questionable.

NUCLEAR STIMULATION—A NECESSITY?

As the study by the Mineral Resources Institute Potential Gas Agency makes clear, many potential new sources of natural gas lie in inaccessible areas, either deep under the ocean floor, or in deep underground rock formations of low permeability and porosity. It is this deep underground potential supply that was thought to be most susceptible to nuclear development under the Plowshare Program, since conventional technology did not recover sufficient quantities of such gas to make the process economically feasible.

A number of conventional technologies have been tried in deep, nonporous sandstone and shale layers such as those that characterize the Gasbuggy, Rulison, and Rio Blanco sites, but they have never met with economic success. Acid leaching is one way of expanding and widening the existing fracture system in deep, tight, nonporous formations. This technology involves the

27

use of a strong acid which is pumped into a well at high pressure. The gas-bearing rock must be open to corrosive attack by the acid for this technique to be successful. Tightly compacted sandstone-shale formations, however, do not have extensive fracture systems and thus are not particularly susceptible to acid leaching.

There is an older technique which involved the use of ordinary high explosives to fracture rock deep underneath the Earth's surface. Technological development of other methods along with the obvious danger of handling high explosives greatly restricted employment of this method, which has not been widely used for the past twenty years. Another drawback involved is that conventional high explosives may not be powerful enough to fracture rock formations of great thickness and low permeability. The basic principle in this technology, however, is basically the same as that involved in using nuclear explosives: the more a dense rock formation is cracked open, and the more extensive the resulting fracture system, the greater will be the amount of gas released. Only the explosives differ. Though cost and safety analyses for the use of nuclear and conventional explosives vary, the basic arguments advanced against the use of ordinary high explosives could also be cited in opposition to the use of nuclear explosives.

Conventional hydraulic fracturing is another technique which has recently been used in tight sandstone and shale formations. Hydraulic fracturing is accomplished by pumping a mixture of liquid and sand under very high pressure into a well drilled in a gas-bearing rock formation. The high pressure fractures the rock and forces a mixture of fluid and sand to flow into the cracks and fissures that are opened. The liquid in the cracks presses on a widening area, while the sand acts as a prop which keeps the fractures open and prevents them from sealing up again under the weight of the earth above. Thus, pathways are opened up which allow the gas held in the formation to flow through the fracture system to the well, where it can then be retrieved at the surface.

Tightly compacted sandstone-shale formations unfortunately have extremely low permeabilities and are often too dense to be penetrated by the hydraulic fracturing method. Although some flow of gas can be stimulated in tight formations by this technology, the amount of natural gas produced may not be sufficient to permit economic recovery at current gas prices. According to Plowshare planners' estimates, a number of conventional wells could not produce an amount of gas equivalent to that which could be tapped from one nuclearly stimulated well in the same tight sandstone field. Nuclear proponents also point out that the very large number of wells required by a conventional technology would necessitate large-scale construction of drilling pads, fences, buildings, and roads, all of which cause adverse environmental impacts.

From the perspective of the AEC, it seemed that the foregoing factors indicated the desirability of undertaking nuclear explorations, and the use of a nuclear technology seemed particularly compelling, in view of the availability of nuclear devices, the prevailing economic conditions, and the national need for more natural gas. The case for trying nuclear stimulation was also fostered in the hope of at last turning the nuclear genie to peaceful and beneficial uses under the Plowshare Program. Much has since been learned about the problems of this technology from the completed Gasbuggy, Rulison, and Rio Blanco tests. If similar projects are undertaken in the future, they conceivably could provide the link between scientific theory and practical application; but the road between the two appears to be heavily strewn with technical, political, economic, and environmental obstacles.

THE WAY IT'S DONE: THE TECHNIQUE OF NUCLEAR STIMULATION

The detonation of a nuclear device underground releases enough energy to melt, crush, and vaporize the surrounding rock. The rock closest to the device is vaporized initially by the blast. Very high temperatures and pressures within the vapor-filled cavity continue the outward expansion in all directions, vaporizing more rock. At the same time, the shock wave generated by the blast continues beyond the edge of the actual cavity, breaking and cracking the surrounding rock, because it cannot withstand the pressure of the wave. Eventually, as the shock wave moves outward, its force diminishes, and the strength of the rock is sufficient to withstand the pressures exerted on it. This outlying rock is "deformed" as the shock wave passes through it, but ultimately it assumes its original position, much as gelatin will quiver, but resettle itself.

Meanwhile, the blast cavity is slowly expanding from the melting of additional rock. As the cavity begins to cool, melted rock collects at the bottom, and within a few hours, the roof of the cavity collapses inward due to fracturing caused by the shock wave's passage. A vertical chimney-shaped hole is formed, which is several times larger than the radius of the original cavity. After cooling, the chimney is composed of three layers: a layer of glasslike melted and resolidified rock at the bottom, an intermediate zone of shattered and crushed rock, and a void space at the top. Figure 2-3 shows the formation of the cavity and the chimney, from detonation to completion of the three-layered configuration.

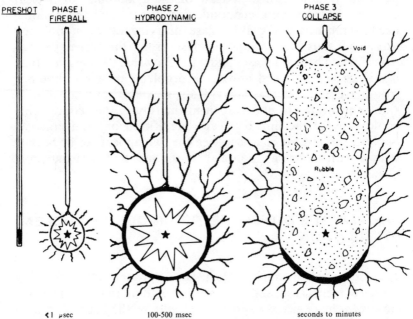

Figure 2-3.   Predicted sequential phases from detonation of nuclear explosive to minutes afterward. (J. W. Watkins, "Fracturing Thick Hydrocarbon Reservoirs with Nuclear Explosives," presented at Southwestern Petroleum Short Course, Texas Tech. University, April 22-23, 1965.)

The chimney and its connected fracture system make up a pathway from which gas can flow into the well. The size of the chimney and the extent of outlying fracturing depend upon the power of the device, the depth of burial, and the geological medium in which the device is detonated.[6]

The rate at which gas can be removed from a well is directly proportional to the extent and openness of the fracture system and the extent of surface area of the cavity. In theory, gas flow is increased in a well stimulated by nuclear energy because the blast creates a larger underground cavity and more extensive fracturing than is possible through conventional techniques. Both the cavity and the fracture system make a larger surface area available for gas transfer. In addition, the cavity serves as a reservoir from which accumulated gas can be brought to the surface.

Figure 2-4 shows the relative surface areas available for gas transfer in wells stimulated by nuclear and conventional means. It can be

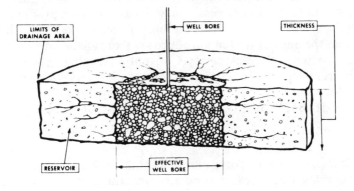

NUCLEAR FRACTURED RESERVOIR MODEL

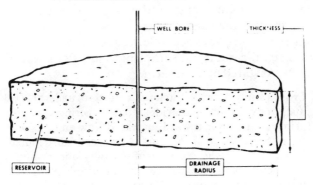

CONVENTIONAL RESERVOIR MODEL

Figure 2-4.   Effect of nuclear fracturing on well bore of reservoir model. (Lynn E. Weaver, ed., *Education for Peaceful Uses of Nuclear Explosives.* Tempe, Arizona: Univ. Arizona Press, 1970, p. 49.)

31

seen that there is a large area of crushed rock and an extensive fracture network in the nuclear model. The conventional reservoir model shows no areas of crushed rock and no fracturing effects, although there would be some fracturing if the well were stimulated by conventional means. As gas is drawn off from a well, the pressure in the hole is reduced; this causes more gas to flow into the hole, and since gas flows faster out of a larger area than a smaller one, the nuclear model would be expected to result in a higher gas-production rate than the conventional model.

It is also well known that shale and sandstone formations are interbedded. The sandstone layers contain gas, but the shale layers do not; the shale prevents the gas from flowing from one sandstone layer to another. The conventional well can therefore produce gas only from the layer in which the bottom of the well is located, whereas the nuclear well can produce gas from all the layers which are connected by the blast cavity. Conventional wells can be opened at different depths, tapping several sandstone layers lying between shale formations, but the total surface for gas to flow would still be expected to be greater in wells stimulated by nuclear explosives.

In summary, three factors potentially contribute to increasing the productivity of a nuclearly stimulated well: the large surface area of the chimney, the fracture system which emanates from the chimney into the surrounding rock and provides a path for gas to flow into the chimney, and the interception of a large number of gas-bearing layers by the chimney and the fracture system. The contribution resulting only from the expectable range of chimney-surface areas, while not significant, is not sufficient to justify the use of nuclear explosives. As stated by Teller et al.:

> From a study of the characteristic effects of underground nuclear explosions, it is evident that fracturing will be the prime contributor toward increased flow and recovery from the reservoir if the fluid is uniformly distributed. In addition to the creation of a cavity and resultant chimney, it is to be expected that a multiplicity of fractures, similar to those resulting from high explosives but of a more extensive and uniform nature, will be developed by a nuclear explosion. This will result in an exceedingly large effective well-bore radius. In addition, bonus

production should be obtained from intervals within the producing formation that are now considered uneconomical to perforate and fracture.[7]

Thus, it is anticipated that for nuclear explosives to be economically useful in gas stimulation, it is necessary that an extensive fracture system beyond the chimney be produced by the explosion and that these fractures remain open for the 10 to 20 productive years of the well.

Assuming that these models are correct and that considerably greater gas production is possible with a nuclear than a conventional technology, it should nevertheless be kept in mind that this comparative advantage must still be weighted against possible adverse factors, such as increased economic, environmental, and social costs.

The AEC has set explicit guidelines for the kind of site that can be chosen for a nuclear stimulation project. These guidelines state that the site should be reasonably remote from habitation, yet easily accessible; that there should have been sufficient drilling in the surrounding area to provide adequate production data, but that the area should not be so highly developed as to incur heavy liability for damage which might be done to existing wells and surface field structures by a nuclear blast. The gas field to be exploited by nuclear technology must be deep, so that the explosion is confined to a specific geological formation, but not so deep as to necessitate excessive drilling, emplacement, and testing expenses. The reservoir layers of rock must be thick and of low permeability and located in areas in which conventional stimulation techniques have proven inadequate. Finally, the AEC guidelines specify a site where there is no subsurface water which could become radioactively contaminated as a result of the detonation and subsequent fracturing. Since no site is ever ideal, it cannot be expected to hold strictly to these guidelines in every respect, but they do indicate the general nature of AEC site standards.

The sequence of steps leading to a nuclear stimulation experiment start with the preparation of a feasibility study and design work for a nuclear well after a site generally meeting the requirements stated above has been found.[8] The feasibility study includes evaluation of the site's characteristics, a proposal suggesting the best type of device to be used, the best depth of burial for the explosive, and the anticipated results from the field.

Site preparation itself starts when the feasibility and design studies have been approved by the AEC and by the commercial sponsors that are involved. The test area is blocked off and cleared for drilling and for installing monitoring equipment, after the necessary permits for land acquisition and use have been secured. A protective building is constructed to house the assembled device before its emplacement. Adjacent areas are cleared for the location of radiological and seismic monitoring equipment. An access road to the site must be constructed if one is not already available. This action alone entails considerable cooperation with concerned state and local agencies, as road building is not a simple administrative matter.

An emplacement hole for a nuclear device is much larger than a hole used in a conventional gas well. In Project Gasbuggy, for example, the down-hole cannister containing the nuclear device had a diameter of 17½ inches and required a correspondingly large well shaft and bore hole. By contrast, the production holes for conventional wells generally are 3 to 5 inches in diameter. During the drilling operations for a nuclear well hole, cables for firing, timing, system monitoring, and electrical safety are laid, which run from the bottom of the hole to ground zero at the surface. The typical casing is made up of sections of steel pipe which are threaded together. As the pipe is lowered into the hole, it is cemented in place against the walls of the well hole. After the cables and the device are lowered into place inside the well casing, the casing must be securely sealed to insure that no radioactivity from the blast will escape to the surface through the emplacement hole.

The nuclear explosive used is not delivered to the assembly building until all site construction, drilling, and laying of cables has been completed. It is checked externally at the site and radiographed to determine if there are any internal defects in the nuclear mass which could cause malfunction. If any component is found defective, a backup device is ready for substitution in place of the original. Control of the immediate environment surrounding the nuclear device is necessary in order to adequately protect it and ensure its reliability. In Rio Blanco, Gasbuggy, and Rulison, high temperatures and pressures underground dictated the use of a corrosion-resistant, pressurized, and refrigerated cannister to provide an adequate environment for the explosive. This type of cannister, which is self-contained and completely sealed off from

34

the surface, has a relatively short lifespan. Once the device is installed, it must be detonated within about two weeks.

The explosive is installed in the cannister in the assembly building, and numerous dry runs are carried out to insure that all electrical, safety, and monitoring systems are functioning properly. Only after the Nevada Operations Office Test Director is satisfied with the results of all preliminary check-outs, are final electrical connections made to the device. The firing cable remains detached from the main connection at the surface and is locked to prevent any possibility of accidental detonation.

About two weeks prior to the date set for detonation, the device is lowered into the casing in the emplacement hold and a bottom-hole check of the electrical system is carried out. If the results are satisfactory, the well hole is stemmed and sealed off. Stemming is accomplished by packing the hole with grout, sand, and pea gravel from bottom to top. The top of the hole is then capped and sealed off. Other wells in the area which could be affected by the blast and subsequent fracturing are also stemmed and capped in a similar manner.

Once emplacement has occurred, the depth and manner of installation together create something like a point of no return. The device either has to be detonated as planned or left in its deep burial ground undetonated in a cannister of a predicted short lifespan. Once a nuclear explosive device has been emplaced, this fact alone militates against any legally imposed short- or long-term postponement of detonation.

When all technical requirements for detonation of the device are complete, the immediate site area is cleared as a safety precaution against possible seismic and radiation hazards. Depending on the kiloton size and the depth of the explosives, it might be necessary to evacuate an area within 5 to 15 miles from surface ground zero. In Projects Rulison and Rio Blanco, safety precautions included disconnecting residential gas lines in the vicinity, rerouting and rescheduling air and rail traffic, and sealing off roads leading to the site. When the test site has been cleared, the explosive is armed by unlocking the firing cable and matching up all final electrical connections. The countdown proceeds then until the device is exploded. Surface ground zero and the surrounding area is reoccupied only after checks for radioactivity and other blast hazards are carried out by hazard-control technicians.

In the face of the AEC's experience and elaborate safety precautions, why is there opposition to the peaceful uses of nuclear explosives? Opponents have raised several arguments. It can be speculated that the very idea of using nuclear explosives—with their roots in the atomic bomb—raises the spectre of a mushroom cloud in many minds. The more tangible grounds of opposition include: effects of radiation exposure, effects of earth motion, and priorities in resource development. Radiation exposure from natural gas stimulation could potentially occur due to accidents in handling the explosive device, accidental venting or leaking from the well, or from contamination of mobile ground water by radioisotopes produced by the explosive. Due to the persistence of some isotopes, such contamination could be of concern for decades and even centuries. Radiation exposure will occur when the gas is brought to the surface, and it would affect large segments of the population if the gas were used commercially.

An underground explosion of the size contemplated for most peaceful applications produces earth motions comparable to an earthquake of average intensity—one that is perceptible and that can cause moderate damage to structures and possibly produce land and rock slides over a distance of several miles. To the senses, this is the most noticeable effect of a nuclear detonation. Closely related is the concern over subsurface damage. The explosion shock might also damage underground geological features. This is especially important when the explosion is in the vicinity of other potentially valuable resources, such as oil shale. It has even been suggested that the shock might trigger other earthquakes.

Still another concern has been over the use of federal funds to promote nuclear technology. Not only are tax revenues used, but the emphasis on a nuclear method might inhibit research and development of alternative technologies. Another argument advanced is that because nuclear explosives have not lived up to expectations, no further time, effort, talent, and money should be invested in this enterprise.

These and other concerns have been voiced, but the central issue that appears over and over again is that of radiation exposure. It is appropriate then to offer here a general explanation of radiation exposure from nuclear stimulation. The focus in later chapters which deal with each specific project is on the reporting of the climate of opinion, whether that opinion was expressed by representatives of governmental agencies, private industry,

36

political leaders and organizations, or independent groups or individuals. Most often, however, the opinions were cast in terms of the general radiological safety of the technology, or conversely, the potential hazard that accrues from its use.

Since the subject of nuclear technology and radiation is an emotional issue for many, it is well to point out that no particular stand is taken here regarding the radiological health hazards involved in nuclear processes. It should be emphasized that nuclear explosives differ in many respects from nuclear reactors and from radioactive isotopes used in medicine. Furthermore, no one should equate the use of nuclear devices in gas stimulation with military uses of nuclear bombs.

The purpose of this analysis is to note some of the merits and the risks of using nuclear underground devices as explosives and to consider the history and future of nuclear stimulation technology as dispassionately and objectively as possible. The technology, as considered here, focuses both on the number and on the scale of actual or contemplated explorations, since the safety or danger from ordinary natural radiation or manmade radioactivity is based on short-terms as well as cumulative dosage. Nuclear power, like any other technological development, is a tool which may be used for good or ill. There is no way to turn back the clock to the pre-atomic age. Our task will therefore be to focus on the relative risks, costs, and benefits of this particular nuclear technology in the light of contemporary human needs.

RADIATION: THE UNKNOWN QUANTITY

Though nuclear stimulation projects have been criticized on many grounds, the chief concern of citizens seems to be the fear of radiation. Part of the concern stems undoubtedly from the ordinary citizen's inability to comprehend the scientific complexity of the information that is available on the effects of radiation. But beyond this, there is such a bewildering diversity of scientific information and opinion as to put the whole question beyond the grasp of most people. Citizens thus do not know what or whom to believe, much less what the results of radiation exposure will mean to them, to their children, and to their future.

The risk of exposure to ionizing radiations have been studied perhaps more thoroughly than any other type of risk exposures. Yet, not all the answers are known. Ionizing radiations cause gene mutations and chromosome aberrations, the effects of which are transmitted to later generations. They also can result in somatic effects; that is, effects on the individual receiving the radiation dose. These include affecting the physical and mental growth and development of youngsters as well as causing diseases such as leukemia and thyroid tumors.[9]

Adverse health effects have been recognized almost since the beginning of man's use of ionizing radiations. As a result, there have evolved organizations and standards for limiting radiation exposures. The early concern was with persons working directly with radiation sources, but increasingly there has been attention to exposure of the general population to radiation.

There are federal radiation standards which limit the permissible human radiation dosage, but the manner in which the standards were set is still argued, and some members of the scientific community remain critical of them.[10] Prior to 1955, there was no government agency which had the responsibility for setting "safe" radiological standards. The National Committee on Radiation Protection and Measurements was composed of a group of private scientists who conducted research and published findings on acceptable health standards, but this group had no official government status. The International Commission on Radiation Standards also conducted radiation-hazard research, but again without official recognition or legal and political authority.

By 1955, the United Nations' Scientific Committee on the Effects of Atomic Radiation was established; in 1958, that committee published the results of a 30-month study by 87 scientists from 15 different countries. The report called attention to the dangers to public health from radioactive fallout from testing of nuclear weapons; it reflected a growing public uneasiness. The large number of nuclear-weapons tests being carried out at the Nevada Test Site made the issue topical and did not lessen public concern. President Eisenhower, by executive order, responded to this increased awareness by creating the Federal Radiation Council (FRC) in August 1959. The FRC was to advise the President on radiation matters affecting the public health and to guide all federal agencies in the formulation of radiation standards.[11]

38

Under the President's Reorganization Plan of 1970, another federal agency has also been given responsibility in the matter of radiation guidelines and standards: The Environmental Protection Agency (EPA) was established, and it set up a Division of Criteria and Standards under its Radiation Office, which took over most of the radiation standard functions of the federal government. The task of the RCD and the newer EPA division has been, and continues to be, to estimate the risks involved in radiation exposure and to balance those against the benefits derived from it. From this kind of evaluation, radiation protection standards are formulated. But the difficulty of the task is illustrated by a Radiation Council Staff Report in 1960 which noted that ". . . reducing risk to zero would virtually eliminate any radiation use and would result in the loss of all possible benefits."[12]

The FRC, and now the EPA, established a Radiation Protection Guide for the general population. According to this guide, the average whole-body exposure should remain below 170 mrem per year[13] (excluding natural background and intentional medical exposures). The mrem (milli-roentgen-equivalent-man) is a basic measure of radiation absorbed and its "biological effectiveness." Thus, when radiation dosages are expressed in mrem, they are reduced to an equivalent basis; that is to say that the effect resulting from a dosage of 100 mrem is presumably the same, regardless of the source or type of radiation involved. This guide is defined as "the radiation dose which should not be exceeded without careful consideration of the reasons for doing so; every effort should be made to encourage the maintenance of radiation doses as far below this guide as practicable."[14] The FRC also indicated that "there should not be any man-made radiation exposure without the expectation of benefit resulting from such exposure."[15] The recommended guide was established on the basis that a

> permissible dose may then be defined as the dose of ionizing radiation that, in the light of present knowledge, is not expected to cause appreciable bodily injury to a person at any time during his lifetime. As used here, 'appreciable bodily injury' means any bodily injury or effect that the average person would regard as being objectionable and/or competent medical authorities would regard as being deleterious to the health and well-being of the individual....[16]

The REC, EPA-recommended guide is official guidance for federal agencies. The guide is the primary basis for all radiation protection standards—all others are based upon and derived from it. Regulations and standards based upon the guide have the force of law when processed through the Administrative Procedures Act.[17]

The route to arrive at a standard for, say, a maximum permissible radioisotope concentration in drinking water, or in natural gas, or the maximum permissible emission from a nuclear power plant, involves many calculations and assumptions.[18] It is a well-developed, but not an exact, science, and controversy and difference of opinion continue over the guide and the manner in which it was arrived at. It is generally agreed that most genetic and many somatic effects of radiation are proportional to the dose—that is, any dose, no matter how small, will produce some effects, and an increase in the dosage increases the effects.[19] Thus, the concept of a "permissible" dose implies an acceptable level of damage. The controversy has two aspects: one is the estimation of the risk associated with exposure, and the other is the acceptable level of effect.

The most recent, detailed study related to establishing radiation protection standards, the BEIR Report, was prepared by a committee of the National Academy of Sciences, submitted to the EPA, and published in November 1972.[20] Although the BEIR Report did not make numerical recommendations, it did state that

> The present guides of 170 mrem/yr grew out of an effort to balance societal needs against genetic risks. It appears that these needs can be met with far lower average exposures and lower genetic and somatic risk than permitted by the current Radiation Protection Guide. To this extent, the current Guide is unnecessarily high.

The report also stated that "the exposures from medical and dental uses should be subject to the same rationale. To the extent that such exposures can be reduced without impairing benefits, they are also unnecessarily high."[21]

Several lines of reasoning have been used in assessing risks. These analyses make use of existing data on human and laboratory animal exposures and information about natural background radiation exposures.

Some perspectives on the guide may be gained by considering natural background. Radiation strikes the earth from space

40

(cosmic radiation) and is received from naturally occurring radioisotopes in the Earth and in all living things, including the human body. In the United States, the average exposure from natural background is about 100 mrem per year, although it varies from about 75 to 170 mrem per year, depending upon location. In mile-high Denver, background radiation is about 170 mrem per year compared to about 75 in sea level Atlantic City. Add to background the average medical dose of about 70 mrem per year in the United States,[22] and it is seen that the recommended Protection Guide is a maximum additional dose about the same as the average already received. As stated in the BEIR Report:

> To summarize: our first recommendation is that the natural background radiation be used as a standard for comparison. If the genetically significant exposure is kept well below this amount, we are assured that the additional consequences will neither differ in kind from those which we have experienced throughout human history nor exceed them in quantity.[23]

Some critics of the guide argue that too little is known, particularly about long-term genetic effects, to allow an additional exposure even equal to background to the general population.[24] Critics also argue that the estimates of risk are far too low based upon the available data.[25] There is, however, almost complete agreement among students of radiation effects that there is some risk associated with any exposure.[26]

Much of the difficulty in assessing general population risks stems from dealing with low levels of radiation dose. Since doses less than about 25,000 mrem produce no immediately detectable clinical effects,[27] it is necessary to make inferences about the delayed effects of lower dosages by comparing the occurrence of various symptoms between exposed and unexposed populations. The closer the dose comes to natural background, the more difficult this becomes. Using somewhat different methods and assumptions, the increase in cancer and leukemia deaths in the United States due to a hypothetical additional exposure of 170 mrem per year (above background and medical) has been variously estimated. Storer's analysis[28] suggests an increase of 160 deaths per year; Gofman and Tamplin,[29] 32,000 per year; Pauling, [30] 96,000 per year; and the BEIR Report,[31] 5,000 to 7,000 per year. These estimates may be compared to the experience of some 310,000 deaths from cancer and leukemia annually in the United

41

States. The great differences between these estimates suggests the difficulty in risk assessment. It should be noted that the estimates are only for cancer and leukemia deaths and do not include any other form of health effect. It should also be emphasized that they are based on an average, total-population exposure of 170 mrem per year, whereas the actual exposures experienced above background and medical are estimated to be of the order of 3 mrem per year.[32]

Just as deriving standards from the basic guide requires assumptions, so do estimates of the doses that might be received from using natural gas stimulated by nuclear explosives. They are just that—estimates. On the basis of the Rulison experience, the estimates predict an average dose increase to users in Los Angeles of less than 5.6 mrem per year.[33] The risks associated with such dose levels can be inferred from the foregoing discussion or by comparisons such as the 70 mrem per year average medical exposure, or the estimate that one cross-country, high-altitude jet flight adds a dose of 3 to 5 mrem.[34]

On the benefit side, the value of having natural gas obtained by nuclear stimulation must be weighed as well as the benefits (and risks) of alternative sources of natural gas or other forms of energy. To date, only very tentative risk-benefit analyses have been attempted.[35]

This discussion may help to put the radiation question in context and perspective, but it does not resolve the issue. It seems that the knowledge that radiation entails risk—plus perhaps an uneasiness with the necessity to judge risk and benefit—generates much of the concern which attends the peaceful uses of nuclear energy.

A PLOWSHARE PAYOFF? SOME GENERAL POLITICAL CONSIDERATIONS

Regardless of these general questions of radiological hazards, technological uncertainty, and economic viability, there are continuing efforts to increase the momentum of the Plowshare Program. The aim has been and continues to proceed from the experimental and developmental stages to actual economic uses of nuclear technology in commercial enterprises. One means has been the introduction of legislation before the Joint Committee on Atomic Energy which would provide a commercial nuclear explosive service available to industry.

42

Another means of converting Plowshare projects into actual commercial use and services would be through the sale and distribution of natural gas that has already been stimulated in Plowshare projects. Sale of gas from Project Gasbuggy is highly unlikely without a significant amount of dilution with nonradioactive gas, and such a course probably would not be worth the additional cost and engineering trouble. The gas from Project Rulison might offer some potential for sale and distribution. If Project Rulison gas can be sold, then gas resulting from Project Rio Blanco could presumably also find its way into a commercial distribution pipeline. This would be the start of the long-sought but so far elusive Plowshare payoff.[36]

At present, there is no provision whatsoever under AEC rules and agreements for the sale of gas released through Plowshare Program experiments. In addition, at this time, Plowshare-stimulated gas cannot be marketed legally because no person or group is authorized to receive it. Only persons and companies specifically licensed by the AEC or an "agreement state" have the power to possess the gas. For example, Austral Oil Company and the Colorado Interstate Gas Corporation cannot sell Project Rulison gas to the Rocky Mountain Gas Company unless the latter obtains a license to receive it. To date, the AEC has quite effectively regulated and controlled possible use of nuclear stimulation experiment natural gas by simply *not* making use of its licensing power.

Speculatively and theoretically, the AEC once could have granted such a license. But it would have had to amend its own administrative regulations as given in Title B, U.S. Federal Code of Regulations. The legal authority for this, then, would fall into the specialized area of administrative code and administrative law. The complexities of large organizational administration also are entailed in the granting of a license. It is also possible that a natural gas distributing company could be exempted from the requirement of obtaining such a license. The AEC could have done this on its own initiative, but would only do so as the result of a legal court petition presented by interested gas companies. In answer to inquiry from the authors, the Austral Oil Company indicated that it had not formally requested AEC action to permit distribution of Project Rulison gas, but that consideration was being given to the matter.[37] Subsequently, there have apparently been some negotiations carried out with a view to eventually distributing and marketing the gas in the Aspen, Colorado area, but these negotiations were prior to the establishment of the Energy Research and Development Administration.

It is not clear at this time whether a producer and distributor applying for a license also would be required to submit an environmental impact statement dealing with the environmental effects of distributing the gas. One representative of a private industry working with the Atomic Energy Commission on nuclear stimulation projects pointed out that meeting the requirements involved in preparing and submitting an environmental statement could cause an additional delay in getting the gas into a pipeline and on its way to consumers.[38] The matter of additional expenses might here also make the whole matter not worth pursuing, especially if protective environmental expenses were added to the enormous initial costs. A further factor complicating the marketing process is the opposition of some citizens in Aspen to the use of Rulison gas in their homes and places of business. As a practical matter, consumer resistence has to be taken into consideration.

Theoretically, the State of Colorado, as an "agreement state," could also issue a license or grant an exemption to a gas company for receiving gas from a nuclear stimulation experiment. This authority lies in the general powers transferred to Colorado by the AEC to regulate by-product materials. The AEC concluded such agreements with the State of Colorado and a number of other states under Section 247 of the amended Atomic Energy Act of 1954. However, even if a gas company was granted a license or exemption by Colorado, it would still be subject to AEC regulatory powers or those of the new Nuclear Regulatory Commission. And additionally, under the AEC Code of Regulations, the transfer of possession or control by the producer of gas containing by-product material to persons or companies exempted from licensing requirements would not alter the fact that the distributor would still need an AEC authorized license to market the gas. The net effect of these provisions is that the Atomic Energy Commission or Nuclear Regulatory Commission retains final legal and practical authority over distribution of Plowshare gas to consumers, whether they live in an agreement or nonagreement state. Hypothetically, the State of Colorado could go further than that. Without subsequent affirmative action or major administrative rule changes by the AEC or its replacement agency, the route from nuclear well to commercial pipeline is nonexistent to date.

It seems very unlikely that the new Nuclear Regulatory Commission will take such action in the near future; practically, if it did, it would be faced with both its own legal and administrative rule tangles and outside opposition. However, if we assume that all of the above hypothetical requirements are met, the Federal Power Commission would then have to make a determination as to whether natural gas produced under the Plowshare Program comes under its rate-making jurisdiction. As a separate governmental agency, the Federal Power Commission has jurisdiction over conventionally stimulated natural gas from production to the setting of the rate schedule and the structure that governs sales and distribution across state lines. The Department of Transportation also has federal authority over pipeline safety, and sets general transportation regulations for interstate shipment of nuclear materials. If either or both of these agencies determined that interstate considerations were involved, the producers and distributors would then have to meet their requirements. The satisfactory fulfillment of Nuclear Regulatory Commission requirements and possible attendant and overlapping Environmental Protection Agency stipulations complete the legal and jurisdictional tangle.

If we assume negotiations for Project Rulison gas between Austral Oil Company and an intrastate marketing firm could survive the hurdles of Nuclear Regulatory Commission and Environmental Protection Agency requirements, the next agency entering the picture is the state regulatory agency—in this case, the Colorado Public Utilities Commission. This commission would assume jurisdiction over gas production and rates, if the gas were to be distributed only within the State of Colorado.

In looking at the overall picture, there does not seem to be much chance that natural gas stimulated by nuclear means will be sold in the near future from the Plowshare projects that have already taken place in Colorado. The legislative and administrative requirements that must be met before the gas can be marketed are only part of a wider governmental and political process. Even if nuclear technology proves adequate to the task of providing a safe and economically viable natural gas product, the basic problem still remains: convincing the general public, governmental agencies, and political leaders that nuclear stimulation projects are in the best interests of all citizens, the state, and the nation.

NOTES

[1]Walter Benjamin, "Future of Energy Sources Bleak," *Boulder Daily Camera,* Nov. 15, 1972.

[2]For example, see Ruben T. C. Guiness, "The Energy Crisis: Real or Imaginary?" *Chemical Engineering Progress,* April 1972; and Environmental Policy Division, Legislative Reference Service, Library of Congress, "The Economy, Energy and the Environment," Sept. 1, 1970; and John J. McKetta, Jr., *Report to the Secretary of the Interior of the Advisory Committee on Energy,* U.S. Dept. of Commerce, Washington, D.C., (PB-201-071), June 30, 1971.

[3]Potential Gas Committee of the Potential Gas Agency, "Future Natural Gas Will Be Expensive To Find," Mineral Resources Institute, Golden, Colorado, June 2, 1971.

[4]"The Energy Crisis," *Rocky Mountain News,* Dec. 1, 1972; and Potential Gas Agency, "Summary Report," 1971.

[5]The General Accounting Office, Study of the Federal Power Commission. Statement issued Sept. 14, 1974.

[6]E. Teller, W. K. Talley, G. H. Higgins, and G. W. Johnson, *The Constructive Uses of Nuclear Explosives* (New York: McGraw-Hill, 1968), Chap. 4.

[7]*Ibid.,* p. 252.

[8]This discussion of the steps and precautions in conducting a nuclear stimulation experiment is adapted from several sources. For a detailed description of the site activities, see, for example, *Project Rio Blanco Definition Plan,* Volume II, CER Geonuclear Corp., Las Vegas, Nevada, March 2, 1972.

[9]A detailed discussion of the effects may be found in the BEIR Report, "The Effects on Populations of Exposures to Low Levels of Ionizing Radiation," *Report of the Advisory Committee on the Biological Effects of Ionizing Radiations,* National Academy of Sciences-National Research Council, Washington, D.C., Nov. 1972, hereafter cited as BEIR Report.

[10]See H. Peter Metzger, *The Atomic Establishment* (New York: Simon and Schuster, 1972), pp. 99-102, for a critical discussion of atomic fallout. Metzger feels the AEC in particular has misjudged and mishandled the matter, but his book considers more than the specific Plowshare Program projects under discussion here, and his criticisms thus are necessarily more general and directed to the entire scope of AEC projects and history. Specific critical articles and books abound on the subject of fallout, and the wealth of explanatory and historical material available on atomic energy is impossible to cite here. Metzger's *The Atomic Establishment* or Glenn T. Seaborg and William R. Corliss's book, *Man and Atom* (New York: Dutton, 1971) are two recent books that are easily available, and they take opposite views on the matter. The latter book is especially good at presenting readable nontechnical explanations of the scientific techniques underlying nuclear technology for the general reader.

[11]Executive Order 10831, Aug. 14, 1959.

[12]Federal Radiation Council, "Background Material for the Development of Radiation Protection Standards," Staff Report No. 1, May 13, 1960, p. 24.

[13]The mrem (milli-roentgen-equivalent-man) is a basic measure of the radiation absorbed and its "biological effectiveness." Thus, when radiation dosages are expresed in mrem, they are reduced to an equivalent basis; that is to say that the effect resulting from a dosage of 100 mrem is presumably the same regardless of the source or type of radiation involved. For a precise definition and discussion of radiation dosage, see K. Z. Morgan and J. E. Turner, eds., *Principles of Radiation Protection* (New York: Wiley and Sons, 1969).

[14]Federal Radiation Council, "Background Material," p. 37.

[15]BEIR Report, p. 8.

[16]Federal Radiation Council, "Background Material," p. 23.

[17]For discussion of this point, see testimony of Paul C. Tompkins, in *Hearings on Underground Uses of Nuclear Energy, Part 2*, Subcommittee on Air and Water Pollution of the Committee on Public Works, U.S. Senate, Aug. 5, 1970, p. 642.

[18]See Morgan and Turner, *Principles of Radiation Protection* for a discussion of the determination of standards and dosages.

[19]BIER Report, Chaps. V, VII.

[20]BIER Report, Nov. 1972.

[21]*Ibid.*, p. 2.

[22]*Ibid.*, pp. 12, 19.

[23]*Ibid.*, p. 52.

[24]For example, see the discussion by J. W. Gofman and A. R. Tamplin, "Nuclear Energy Programs and the Public Health," *Hearings on Underground Uses of Nuclear Energy*, Appendix A, p. 1377.

[25]J. W. Gofman and A. R. Tamplin, *Poisoned Power* (Emmaus, Pa.: Rodale Press, 1971), passim.

[26]See BEIR Report, and the statement by K. Z. Morgan in Hearings on Underground Uses of Nuclear Energy, p. 643.

[27]Federal Radiation Council, "Background Material," p. 5.

[28]J. Storer, "Memorandum," in *Hearings on Underground Uses of Nuclear Energy*, p. 1384.

[29]Gofman and Tamplin, *ibid.*, p. 1377.

[30]L. Pauling, *ibid.*, p. 1362.

[31]BEIR Report, p. 91.

[32]*Ibid.*, p. 19.

[33]G. C. Werth, M. D. Nordyke, L. B. Ballou, D. N. Montan, and L. L. Schwartz, "An Analysis of Nuclear-Explosive Gas Stimulation and the Program Required for Its Development," Lawrence Radiation Laboratory Report (UCRL-50966), 1971.

[34]Statement of C. E. Larson in *Hearings on Underground Uses of Nuclear Energy, op. cit.*, p. 672.

[35]H. J. Otway, ed., "Risk vs. Benefit: Solution or Dream, an Informal Report," Los Alamos Scientific Laboratory (LA-4860-MS), 1972.

[36]For discussion of the developments to date for sale and commercial use of gas stimulated by nuclear means from Project Rulison and Rio Blanco, see Chapters 6 and 7.

[37]Letter from Mr. Miles Reynolds, Jr., Vice-President of Austral Oil Company, to Catherine Wrenn, July 14, 1972, University of Colorado, Boulder, Colorado.

[38]Conversation with Mr. Harold Aronson, Vice-President of CER Geonuclear Corporation; Catherine Wrenn and Betty Arkell interviewed Mr. Aronson on July 11, 1972, in Boulder, Colorado.

Chapter
3
The Course of Project Gasbuggy

The Gasbuggy shot to explore the feasibility of nuclear stimulation of natural gas was detonated December 10, 1967, in New Mexico. A few days later, the *Farmington Daily Times* called this event one of the "top state stories" of the year.[1] This designation was not surprising, because the Gasbuggy detonation had been heralded by the New Mexico Governor, the State's Senators, and members of the Joint Committee on Atomic Energy as an event "of tremendous and incalculable profit to the people of the United States."[2]

Although the Atomic Energy Commission's Plowshare Program had felt a financial pinch in 1967, the segment of the program devoted to nuclear stimulation of natural gas moved steadily ahead, sustained in large part by the widely predicted nationwide shortage of natural gas. Gasbuggy's sponsors put forward a number of reasons for their project, but underlying the arguments was the contention that it was a paramount national priority to supplement the nation's rapidly dwindling gas supplies. This increasingly obvious national need worked to the advantage of the AEC and the El Paso Natural Gas Company, the Project's industrial sponsor, in gaining the kind of legislative and administrative decisions necessary to fund the Gasbuggy Project. The total estimated cost of $4,700,000 for the project seemed little enough to spend for a program which held such promise for the future. Of the total, El Paso's contribution was to be $1,800,000, and the AEC's was to be $2,800,000. The New Mexico Highway Department and other governmental agencies accounted for the remainder by supplying construction and other services.[3]

The AEC, El Paso Natural Gas, and the United States Department of the Interior (through the Bureau of Mines) supported the case for jointly carrying out the project:

> Proof of the utility of NE [nuclear energy] can only be determined by a full-scale field test. Justification for such a test lies in the number and the real extent of low productivity reservoirs where the ultimate recovery and the rate at which such recovery occurs may be substantially increased.[4]

The history of the Gasbuggy experiment began in 1958, when El Paso Natural Gas Company initiated correspondence with the Lawrence Radiation Laboratory of the University of California about the feasibility of nuclear stimulation of suitable reservoirs, which were believed to exist in company-held oil and gas lease holds. Initially, El Paso was interested in a 92,000-acre site known as the Pinedale Unit area in Wyoming. Later, the choice of optimum location for a field test was changed to El Paso leaseholds in the San Francisco Operations Office prepared the initial Gasbuggy feasibility study, and El Paso and the Bureau of Mines provided supporting industrial-technical data and engineering evaluations. The preliminary design of the experiment called for the drilling and casing of an emplacement hole to house a 10-kiloton nuclear device just below the base of the Pictured Cliffs geological formation. The force of the nuclear blast was expected to be 5,000 times greater than that generated by the 1,000 quarts of solidified nitroglycerin which had previously been used to stimulate wells in this area. Subsequent analytical study, however, apparently convinced project planners that a 10-kiloton device would not generate a force of sufficient magnitude. It was ultimately decided to use a 26-kiloton device.

The sponsors believed that only through a full-scale field test could the answers to certain fundamental questions about nuclear stimulation of natural gas be answered. Through the Gasbuggy Project they hoped to achieve the following goals:

> 1. Determine the differential in productivity of a nuclear as opposed to a conventionally stimulated well and then to be in a position to effectively estimate the in-

crease in ultimate productivity resulting from nuclear stimulation.

2. Obtain data on the quality of the nuclearly stimulated gas in respect to residual radioactivity and to evaluate various purification techniques. These data would then be used to determine if the resulting gas would require treatment before it could be utilized commercially.

3. Develop engineering knowledge and capability necessary for future nuclear gas-stimulation projects.

4. Refine existing predictive capability as to seismic effects and ground motion on gas field equipment and such other structures as might be adversely affected by the blast.

5. Obtain data and test predictive capability as to the effect of nuclear blasting on sandstone, including such physical effects as chimney size and extent of fracturing.[5]

Geological studies indicated some particularly appropriate characteristics of the San Juan Basin for an experiment of this nature. First of all, there were known gas reserves in the tightly compacted sandstone of the Pictured Cliffs Formation. Moreover, preshot test wells did not reveal any mobile water which could be radioactively contaminated in the formations which might be affected by the test. Since the closest existing water wells from the nuclear detonation point were located 50 miles away in the Ojo Alamo Formation and 1,700 feet above the detonation point, the sponsors concluded that there would be no ground-water contamination from the proposed blast, but they added that additional wells would be used to confirm these conditions.

The experiment was planned for complete containment of radioactive material underground to avoid atmospheric radioactive contamination from the blast. Nevertheless, the United States Weather Bureau prepared a fallout pattern based upon a maximum credible mishap which could vent radioactive material into the atmosphere. The United States Public Health Service did a preliminary survey within a 100-mile radius of the test site and concluded that adequate precautions could be taken should such an unforeseen venting occur. This conclusion was at least in part based on the fact that the area surrounding the test site was relatively free of milk cows. If permitted to forage on grasses affected by radioactive fallout, cows could pass along contamination to humans through their milk. In addition, the sponsors noted that the few people who lived in the area were concentrated in small population centers which could be readily monitored and controlled, particularly as the Project planners had ". . . good working relationships with public officials of the State of New Mexico."[6] Figure 3-1 illustrates the relative isolation of the test-site area.

Additional surveys of the area showed that the only structures close enough to ground zero to conceivably be damaged by the shock effect were owned by El Paso, including only one existing gas well. It was suggested that if any damage were done to this well, it would provide experimental data on the kind of shock damage to gas field installations which might be anticipated in future nuclear stimulation projects.

After the AEC and its cosponsor were convinced of the merit and safety of the project, they took steps to gain public acceptance of their proposed venture. The Gasbuggy Public Information and Observer Program was created to provide an "information package [and] integrally with this to disseminate full information on a timely basis to promote both the use and understanding of peaceful nuclear explosions."[7]

To achieve these goals, the sponsors agreed to provide, both independently and through joint action, information on the project to newspapers, radio and TV networks, and wire services. Particular efforts were made to disseminate information to residents of the El Paso Natural Gas Company consumer area, but material was also distributed to local, state, and congressional officials, federal officials, and both domestic and foreign industrial groups potentially interested in the project. The AEC, the Bureau of Mines, and El Paso were responsible for disseminating information within their own particular sphere of influence, as well as for cooperating with the other partners in matters of mutual concern.

Public information efforts ranged from news releases to radio tapes, TV clips, still photography, film footage, and brochures. In addition, El Paso ran full-page newspaper ads several days prior to detonation entitled "Project Gasbuggy: A Dramatic Experiment Which May Lead to a Major Breakthrough in Natural Gas Production."[8] Through graphic representations, the

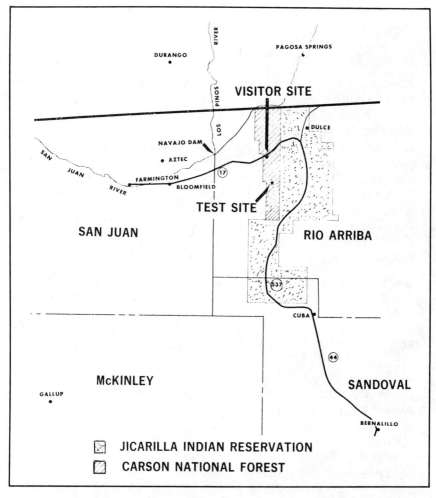

Figure 3-1. Test site area. (*Project Gasbuggy*, Joint Office of Information (PNE-G-4), Sept. 15, 1967.)

53

ad effectively explained the rudiments of nuclear stimulation to produce well fracturing in tightly compacted formations, and it included the Bureau of Mines estimate that "recoverable reserves [of natural gas] of the nation could be more than doubled" through this process.

Newspaper coverage was extensive several months prior to the blast in both the *Albuquerque Tribune* and the *Albuquerque Journal* and in the *Farmington Daily Times*, which serves the area closest to the Gasbuggy site. For more than a year preceding the blast, there were periodic articles explaining the nature and purpose of the project and its progress to date. The AEC, in conjunction with the New Mexico Health and Welfare Department, distributed a pamphlet, printed in English and Spanish, throughout northwestern New Mexico. The pamphlet, explaining the potential of the Project for stimulating economic growth and social progress throughout the area, was posted in prominent places and distributed to schools in the area.[9]

A review of the newspaper coverage shows that the press took a very favorable position toward the project. There was little or no public reaction which would cause them to report unfavorable comment; public officials, both in and out of the State of New Mexico, voiced strong support for the project. As early as September 1966, Senator Peter Dominick of Colorado pressed for Senate approval of $750,000 "to free huge pockets of natural gas in northern New Mexico through nuclear explosions." He cited the tremendous economic potential of this kind of recovery technique, adding that "the project could exceed our most optimistic expectations."[10] New Mexico congressmen consistently pressed for progress on Gasbuggy, and some were unhappy with the AEC for what they felt were unwarranted delays in the Gasbuggy timetable. Congressman Thomas G. Morris, a member of the Joint Committee on Atomic Energy, in an interview with the *Farmington Daily Times*, said that in his opinion, "the AEC hasn't helped us a damn bit with Project Gasbuggy."[11] Later at predetonation ceremonies, he expressed admiration for El Paso Natural Gas Company, which "had the guts to put their money where their mouth is."[12] Senators Clinton Anderson and Joseph Montoya of New Mexico supported the project as consistently, though perhaps not as vehemently. Their evident support for the project was shared by New Mexico Governor David Cargo, who requested and got from the state legislature authorization to build a $100,000, 5.3-mile, all-weather road to serve the Gasbuggy site.[13]

The Jicarilla Apache Indians, whose tribal capital, Dulce, is located about 20 miles from the Gasbuggy detonation point, were apparently not unduly alarmed about the proposed use of their tribal lands. According to Charlie Vigil, the Chairman of the Tribal Council, there was some concern among his people about possible damage to a fault underneath the village which had opened a few inches on the surface during a 1966 earthquake. But Vigil, through the tribal newspaper, spoke reassuringly of the project, noting that "its effects will be hardly noticeable." The Apache Nation's main source of revenue, approximately $1.5 million per year, comes from oil and gas leaseholds on tribal lands, and the Jicarillas were generally not inclined to protest Gasbuggy plans.[14] The day after the detonation, the *Albuquerque Journal* emphasized the concept of Gasbuggy as a boon to the Indian by printing a picture entitled "Space Age First Helps First American,"[15] showing a tribal member with a Gasbuggy crewman.

The day before detonation, the AEC held a Gasbuggy briefing program in Albuquerque for some 275 top-level executives from industry and government. Members of the Joint Committee on Atomic Energy and representatives of the AEC, El Paso, and the Bureau of Mines spoke after a welcoming address by Lieutenant Governor E. Lee Francis. The AEC arranged plans for observation of the detonation and chartered buses to take official visitors to a specially designated observation site. Visitor sites, open to the public, were set up at several locations at roadblocks along the highway.[16]

The actual blast caused some ground motion and a muffled rumble which reportedly awed some observers. The fact that the long distance telephone system at the site was knocked out for seven minutes also caused some consternation.[17] Measurements at the Navajo Dam, 24 miles away, indicated a substantial shock of 4.5 to 5 on the Richter scale.[18] Windows were reported to have rattled in Farmington, 55 miles away.

Half an hour after detonation, a public information officer confirmed a successful detonation and reported no release of radiation and no reports of damage from ground shock.[19] A few hours later, it was announced that a very small amount of radioactivity had leaked out through a cable in the emplacement

55

well hole, but it was quickly sealed off.[20] A water leak into the wall had occurred during drilling and was never completely stopped. The plan had been to detonate in a dry hole, but the experiment proceeded despite the water. The fact that the device detonated and reached full power in a wet hole was considered an advance in the technology.[21]

The initial press statement emphasized that several months of testing and analysis would be required to evaluate the experiment. Only then would the rate and quality of the gas flow and the need for purification techniques become known. The following section sketches briefly the technical course of Gasbuggy and provides some indication of its end result. Though apparently not an unqualified success in fulfilling all of the sponsor's objectives, Gasbuggy was nevertheless a stepping stone on the road to later, bigger, and more ambitious nuclear-stimulation experiments.

A NEW TECHNOLOGY EVALUATED

As finally designed and implemented, the Gasbuggy experiment made use of a nominal 26-kiloton nuclear explosive device. The after-the-fact energy was calculated to have been within the range of 29 ± 3 kilotons. Although the exact nature of the device was then classified because of its similarity to nuclear weapons, it could be deduced that a fusion (hydrogen) device was used,[22] and subsequently it has been reported that in fact this was the case.[23]

There are two types of nuclear explosives, fission and fusion. The type used can be quite important to the results of a gas-stimulation project. A fission explosive releases energy when uranium or plutonium atoms split (hence, fission). Many radioactive isotopes, the majority of which tend to remain as solids, are produced by a fission explosion. The principal, long-lived, gaseous radioisotope produced is krypton-85. A fusion explosive (also known as a thermonuclear or a hydrogen device) is one in which atoms of hydrogen isotopes join (hence, fusion) to make a helium atom and release energy—much as in the sun. A fusion explosive requires a smaller fission device as a trigger. A fusion explosive produces smaller quantities of the heavy radioisotope products characteristic of a fission device of the same explosive energy, but it produces much more tritium, the radioisotope of hydrogen. Tritium is of special concern because it can be incorporated into gaseous products, including natural gas and steam, and because as water it can be transmitted through the food chain and ingested by humans. Thus, tritium is probably the principal product of radiological concern in nuclear gas stimulation. Indeed, it was recognized that applications to hydrocarbon resources such as gas stimulation might require all-fission devices to minimize tritium production. On the other hand, for the same energy yield, a fusion explosive would be less expensive than an all-fission one.[24] Ironically, when a detonation takes place in the atmosphere, fusion devices are often referred to as "clean" and fission as "dirty" because the latter result in much more fallout of radioisotopes.

The Gasbuggy detonation occurred 4,240 feet below the Earth's surface and 40 feet below the gas-bearing Pictured Cliffs sandstone formation. The blast was expected to create a chimney, formed of broken rock, about 160 feet in diameter and 350 feet in height above the detonation point, with fractures extending out 425 feet from the detonation point.[25] These predictions were based on previous underground blasts, most of which were carried out at the AEC's Nevada Test Site. The actual size of the chimney was calculated to be very close to that predicted, about 160 feet in diameter and 333 feet in height, with fracturing extending less than 400 feet from the point of detonation[26] and probably less than 300 feet.[27] The chimney and fracture radii are estimates based on production performance data, since no direct measurements were made. Although measurement of the chimney radius was originally planned, "money limitations forced its elimination."[28] Figure 3-2 shows predicted underground effects of the Gasbuggy shot.

About a month after detonation, the chimney was entered by drilling back through the stemmed emplacement hole. This new hole, the Gasbuggy production well, was labeled "GB-ER." The time span between detonation and reentry was required for drilling the new well, and it also allowed time for short-lived radioisotopes to decay.

The evaluation of gas-well productivity is not a simple matter. A well must be producing for a period of time (several months in the Gasbuggy area)[29] to obtain sufficient information to estimate its

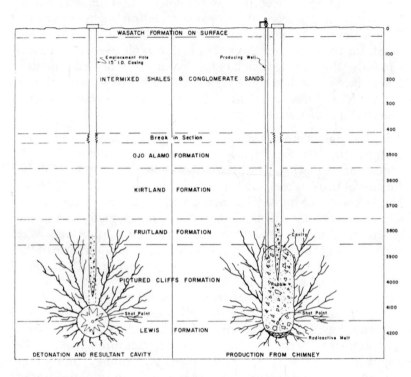

Figure 3-2.   Project Gasbuggy predicted underground effects. (El Paso Natural Gas Co.; The United States Atomic Energy Commission; The United States Bureau of Mines; Lawrence Radiation Laboratory, *Project Gasbuggy*, (PNE-G-1), El Paso, Texas, May 14, 1965.)

properties and to project its long-term productivity. This was done with the Gasbuggy well. The well was also compared with existing nearby wells to determine the extent to which the nuclear explosion had affected productivity.

Production testing included withdrawal of 284 million standard cubic feet of gas from well GB-ER in several stages over a period of almost two years. As is typical of gas wells, the initial productivity was higher than in the later stages. From the later, longer tests, it was estimated that the well could produce 900 million cubic feet of gas over a 20-year productive life.[30] Two wells, GB-1 and GB-2, were drilled near the emplacement well prior to detonation to obtain productivity data under preshot conditions. These two wells were completed without any form of stimulation, and only GB-1 had sufficient productivity to give meaningful results. A 30-day production test of GB-1 gave a flow of 125 thousand cubic feet per day as compared to 930 thousand cubic feet per day for the Gasbuggy well after the same period of production. Since these wells were close together (less than 300 feet apart), this comparison suggests that the nuclearly stimulated well is about 7.4 times as productive as the unstimulated one. A similar comparison with the closest field well, No. 10, which was 435 feet away, which had been producing for several years, suggested a sixfold increase in productivity in the Gasbuggy well.[31]

Gasbuggy's productivity was also compared with five of the eight nearest wells in the field. These wells had been hydraulically fractured. The three most productive nearby wells were not included, "because they [were] further from GB-ER; also, natural fracture systems encountered in these three wells gave them a much higher productivity than is believed representative of the test area."[32] Gasbuggy produced at 6 to 7 times the rate of the average of those five wells. If the other three had been included in the comparison, the increases would have been only a factor of 2.

Some more detailed comparisons are shown in Table 3-1. Note especially that the projected productivity of the Gasbuggy well, GB-ER, was only 2.6 times that of the average of all wells in the field, a figure which is no better than the best conventional well in the field. The results in Table 3-1 emphasize that nuclear stimulation did result in an increased productivity but that the apparent increase is highly dependent upon how the comparison

TABLE 3-1
GASBUGGY PRODUCTION COMPARISON DATA

	Distance to GB-ER (ft)	Daily average production (1,000 ft3/day)[a]	Estimated production in 20 years (1,000,000 ft3)[b]	Estimate of gas recovery in 20 years (percent of gas in 160 acres)[c]	Ratio of estimated 20-year Gasbuggy production to well production
GB-ER	—	135[d]	900	19	—
Well no. 10	435	22	170	3.5	5.3
Average of five nearest wells	< 4500	18	112	2.5	8
Average of all wells in field	< 6800	61	342	7.2	2.6
Best in field	6800	159	900	19[e]	1.0
Worst in field	4500	10.5	65	1.4	14

[a]Based on all production since well was drilled until 1966.

[b]Assuming production in 1966 can be maintained for remainder of the time.

[c]Assuming the same amount of gas in place as at the Gasbuggy site.

[d]Calculated against 500 psi line pressure.

[e]Since this well is completed in a fracture system, this is probably not a valid number.

Source: A. Holzer, *Gasbuggy in Perspective*, Lawrence Livermore Laboratory Report (UCRL-72175). (Last column added by authors.)

is made. Discussions of the results of Project Gasbuggy have consistently referred to a productivity increase by a factor of at least 5 over conventional wells,[33] even though Holzer cautioned that "care should be taken in using the ratio . . . as indications of a figure of merit of Gasbuggy . . . ."[34]

An early prediction that a 10-kiloton explosion was expected to lead to a 20-year production of 3,520 million cubic feet of gas has been reported.[35] An explosive almost three times that powerful resulted in only one-fourth as much gas, and so clearly this early estimate was highly optimistic. Also, clearly, this was recognized before the final planning was done, for it was decided to use the larger explosive.

The extent of fracturing was estimated by observation of down-hole pressures and physical characteristics of nearby wells. Two wells within 300 feet of the detonation point "showed a lack of open fracture communication with the chimney."[36] The extent of fracturing and the increase in formation permeability resulting therefrom are extremely important to the success of nuclear stimulation technology, as has been discussed earlier. A. Holzer, in his summary, *Gasbuggy in Perspective*, stated concerning fracturing, "One of the disappointments of Gasbuggy was that no postshot permeabilities in the 100 to 1000 millidarcy region (similar to those found in the Hardhat measurements) were observed."[37] Project Hardhat was an earlier experiment in granite in which large increases in formation permeabilities outside the chimney had been observed.

Werth *et al.* noted that "Data from Gasbuggy indicate that the fairly high clay content in the gas-bearing Pictured Cliffs Formation acted to inhibit the formation of permanent fractures."[38] Holzer added, "It is rather interesting to note that the flow rate factors of 5 to 8 in Gasbuggy production, as compared to well no. 10, are in agreement with the logarithm of the ratio of the two well bore radii, neglecting any effect due to the hydrofracking of the conventional well. This implies that the ratios observed may be explained by a simple well bore effect."[39] The phrase, "a simple well bore effect" refers to the fact that gas will flow faster into a large area hole than into a small one. The significance of this observation is that the increased gas productivity observed may be due to the chimney alone, with no important contribution from fractures extending outward from the chimney. Indeed, later microscopic examination of samples taken from a well drilled to within 198 feet of the detonation point showed essentially no fractures in the gas-bearing Pictured Cliffs Formation, even though some fractures were found more than 700 feet away in the Ojo Alamo Formation.[40] Thus the Gasbuggy results indicate that, in this field at least, the nuclear fracturing mechanism did not function as anticipated.

The fuel quality of the gas produced from GB-ER was distinctly less than that of the gas in the field before the blast. Early samples of gas contained 36 percent carbon dioxide, which does not burn, an "unexpectedly high" value.[41] They also contained 17 percent hydrogen and 4 percent carbon monoxide, both of which burn, but with substantially less heat per unit volume (as natural gas heating values are measured) than methane and other hydrocarbons. This gas contained a total of only 41.2 percent hydrocarbons, as compared to over 99 percent in the field gas. The "unexpectedly high" amount of carbon dioxide was thought to be due to decomposition of carbonate rocks at the extremely high temperatures of the explosion.[42]

As gas was produced from the well, the carbon dioxide, hydrogen and carbon monoxide contents dropped, the hydrocarbon content rose, and the gas composition began to approach that of the field. This is summarized in Table 3-2. Interestingly though, even in November of 1969, after nearly one-third of the estimated 20-year production of the well had been removed in the tests, the gas had still not quite returned to its predetonation composition or heating value. Of the total gas produced, only three-quarters was hydrocarbons, and its average heating value was only about 80 percent of that of the field gas. This gas composition problem could be overcome in several ways, but it does discount the fuel value of the gas produced, at least in the early stages.

Although shock-wave and ground motion from the project differed significantly from the predicted values,[43] this was not unexpected, because the predictions were based on test results of blasts in different geological media which were carried out at the Nevada Test Site. There were three complaints of structural damage resulting from the blast; only one, at a ranch 25 miles away, was found to be valid.[44] The lack of damage was due primarily to the isolation of the Gasbuggy site. When well No. 10, 435 feet from GB-ER, was reentered, it was discovered that there

TABLE 3-2
GAS COMPOSITION FROM THE GASBUGGY WELL

	Pre-shot	Short-term production test, June 1968	Prior to first 30-day production test, November '68	Prior to extended production test, February '69	End of extended production test, November '69
Percent carbon dioxide	0.29	35.60	24.27	16.37	8.89
Percent hydrogen	—	12.03	10.11	6.38	2.35
Percent methane	85.36	45.45	56.35	65.66	73.24
Percent other hydrocarbons	13.76	6.30	8.45	10.90	14.96
Percent nitrogen and hydrogen sulfide	0.59	0.62	0.82	0.69	0.56
Heating value, Btu per cubic foot	1178	588	790	938	1112

Source: C. H. Atkinson, Don C. Ward, and R. F. Lemon, *Gasbuggy Reservoir Evaluation — 1969 Report*, PNE-G-54, p. 7.

was casing damage 628 feet above the nuclear detonation point and at a slant height of 760 feet.[45]

The Gasbuggy shot did provide new seismic data on the effects of nuclear blasting in deep, gas-bearing formations, since prior AEC tests had been carried out in basically different geological media. These new data were used later as the basis for predicting ground motion and damage from Project Rulison. The possibility that the shot might trigger earthquakes in the local area was apparently ruled out by the Gasbuggy sponsors. No specific provisions were made for evacuating local residents at the Indian tribal center of Dulce. The few Dulce residents who did express concern about renewed earthquake activity were simply told that the shock effects would be hardly noticeable. In fact, the blast-created shock wave *was* hardly noticeable, and no earthquake activity occurred.

Although there is no serious concern about the long-term effects of the seismic activity created by the Gasbuggy experiment, there is, as noted previously, uncertainty about the effects of radioactivity produced by underground nuclear stimulation technology. This issue, which is of major interest and concern to the public, is a most difficult one to discuss in comprehensible terms. An effort is made below to examine, briefly and necessarily superficially, the radioactive products of the Gasbuggy shot and to indicate some implications of the possible use of the gas produced by it.

It is estimated that the Gasbuggy detonation produced 4 grams of tritium, or 45,000 curies, as it is measured in radiological terms, and 350 curies of krypton-85.[46] Since tritium, although radioactive, is chemically identical to hydrogen, it can replace hydrogen in methane and in water and make them radioactive. A large fraction of the tritium in the nuclearly stimulated gas (excluding water) was in the form of hydrogen gas shortly after detonation. At chimney pressures and temperatures, much of the gaseous tritium was incorporated into methane through chemical reactions by the time production testing began.[47] Most of the total tritium was in the form of water, some of it as vapor in the gas, but most of it as liquid in the chimney.

Accurate measurements of the amount of tritium generated by the blast could not be obtained, because water from the Ojo Alamo Formation had seeped into the chimney from the leak mentioned earlier, but it was estimated that only about 5 percent of the total tritium was in the gaseous product. A significant portion of the chimney, perhaps 10 percent, was filled with water from the leak.[48] Since the water coming from the Ojo Alamo Formation was uncontaminated, it diluted the blast-produced water, causing measurements of tritiated water to reflect only a lower limit of contamination. Even so, a tritium concentration greater than one million pCi/cc[49] (pico-curies, a measure of the quantity of radioactivity per cubic centimeter) was measured in early samples. This concentration is more than one thousand times the recommended guideline concentration for drinking water. The importance of proper disposal of any liquid water obtained while producing or processing the gas is quite evident.

An examination of gas samples was made soon after detonation to determine if there were any heavy nuclides (such as strontium-90) in them. The measurements indicated no cause for concern in this respect, for "the limits established . . . are low enough so that the possible presence in the gas of the 'semi-volatile' nuclides can be ignored in the case of Gasbuggy."[50]

The gas produced from the well (excluding water) initially contained tritium at a concentration of about 700 pCi/cc and krypton-85 at about 110 pCi/cc. These concentrations decreased as gas was removed from the well because it mixed with uncontaminated gas flowing into the chimney from the surrounding formation. At the end of the production tests, the concentrations were about 35 pCi/cc of tritium and 7 of krypton-85.[51] Virtually all of the krypton made by the explosive had been removed from the well by then, but most of the tritium was still in the chimney as liquid water. As a basis for some perspective, these concentrations may be compared with the recommended guidelines for occupational exposure for a 168-hour week of 2 pCi/cc for tritiated water vapor and 3 for krypton-85.[52] The guideline values are for air—there are no established standards for natural gas—but this comparison suggests that with the concentrations at the end of the test, about a seventeenfold dilution would have to occur by the time these materials reached the open atmosphere in order to meet the occupational guidelines. Very large dilutions would in fact occur by mixing with air if the gas were burned and the products discharged into the atmosphere or even inside a home.

If it is assumed that Gasbuggy gas were to be used commercially, the two nuclides, tritium and krypton-85, would be the primary causes of radiation exposure to humans. If the gas were burned, all the tritium in the hydrocarbons would be converted to tritiated water vapor (water with one hydrogen replaced by tritium). Individuals could be exposed to radiation through inhalation and absorption through the skin of tritiated water vapor and through immersion in gas containing water vapor and krypton-85. Krypton-85 is an inert gas, not absorbed into body tissue, and so it is not as serious a problem as tritium. It was estimated that 95 to 97 percent of these nuclides present in the gas would remain in the gas after processing and would reach domestic consumers.[53]

There have been several hypothetical studies of the radiation dosage to users of the Gasbuggy gas. The estimates involve several factors, and they differ depending upon the specific assumptions made. Factors which influence the quantity of radioactivity available for potential exposure to domestic users include:

1. concentration of the radionuclides at the wellhead;
2. pipeline dilution;
3. quantity of gas consumed;
4. fraction of combustion products vented inside home;
5. home dilution; and
6. other uses of gas, such as in electric-power plants.

Age, sex, and the amount of time spent in the home would also affect the dose received by an individual.[54]

An early study has been made, giving estimates for home use of the gas in the San Juan Basin based on the tritium content, 420 pCi/cc, in October 1968. It was estimated that in the collection and processing system, the gas would be diluted by gas from other wells in the area in the ratio of 1 part Gasbuggy gas to 49 parts of noncontaminated gas. Assumptions were made for the other factors mentioned above. The calculations were based on unvented heating and appliances, which is not representative of most situations but is a "worst" condition for the amount of combustion products vented inside the home. These calculations resulted in an estimated dosage of about 14 mrem per person per year.[55] A later estimate, for the Los Angeles Basin, gave much lower dosages, mainly because of greater dilutions with uncontaminated gas and lower heating requirements. The maximum individual dosage in Los Angeles was estimated to be 2.2 mrem per year, even when contaminated gas was used in power plants as well as in homes.[56]

The estimated exposure value of 14 mrem per year, while only 8 percent of the recommended dosage guideline of 170 mrem per year for the average exposure per person in a population, may not be insignificant. Most of the assumptions made in this calculation probably tend to overestimate the dose. However, it should be noted that if the Gasbuggy gas was used *without dilution* by uncontaminated gas, even with the tritium concentration of 35 pCi/cc at the end of the test and with all the other assumptions the same as above, the estimated dosage would be about 60 mrem per year. Using a concentration of 100 pCi/cc the estimated dosage would equal the dosage guideline. The gas from the one Gasbuggy experimental well would never be used without dilution, but if a large field were to be developed by nuclear stimulation, this situation could conceivably occur. Whether due to such considerations or others, it was recognized that, "It is pretty clear . . . that these concentrations [in the Gasbuggy gas], especially of tritium, need to be reduced."[57] It became a goal of the Plowshare Program to develop explosives for gas stimulation which would minimize tritium yields.[58]

Holzer pointed out that "Gasbuggy could never be economic from the standpoint of the value of the gas produced, nor was it ever meant to be economic."[59] The project was an experiment to learn about the technology. Its total cost was at least $4,700,000. The total value of the gas which it is estimated could be produced from the well was about $135,000 (at 15 cents per thousand cubic feet, the prevailing price in northern New Mexico in 1967). At a more realistic current figure of 45 cents per thousand cubic feet, the value would be $400,000. This latter figure would just equal the projected cost of the nuclear explosive used.[60] Even though Gasbuggy was not intended to be economic, the results showed that substantial changes in the technology would be required before the technique could possibly become economic. It was evident that smaller diameter explosives would be required to minimize drilling costs.[61] The concept of detonating several explosives, larger than Gasbuggy's, in one emplacement hole, began to be emphasized.[62] Further experiments to develop the technique had been planned, and Gasbuggy pointed to some of the developments which would be needed.

GASBUGGY REVISITED

Project Gasbuggy has been called a "technical success" and characterized as "highly successful."[63] In his review, A. Holzer was perhaps more realistic in stating, "It is my opinion that Gasbuggy as an experiment was eminently successful,"[64] with the qualification that "while the Gasbuggy experiment provided technical information on mechanical effects, gas flow, reservoir evaluation, and radioactivity and chemical composition, the gas reservoir was not sufficiently thick to determine the economic potential of this technology."[65] It seems significant that there have been no further nuclear gas stimulations in the San Juan Basin in the seven years since the Gasbuggy Project was carried out. The El Paso Natural Gas Company had plans for Wagon Wheel, a similar but more ambitious project in Wyoming in 1973 or 1974, but this proposal has encountered a number of obstacles, both technical and political. All things considered, it seems unlikely that an industrial sponsor would delay so long in following up if the first experiment had been clearly successful. It appears that Gasbuggy demonstrated that the technique was not economically attractive in the San Juan Basin.

The shot stimulated gas flow into the well to a degree somewhat greater than had been possible through conventional techniques, but uncertainty remained as to how much improvement had occurred. Seismic measurements indicated that significant damage might result if more powerful devices were used in areas more densely populated and developed than the San Juan Basin. The gas was found to contain surprisingly high concentrations of nonburnable and low heating value gases. The tritium concentration was high enough that reducing it became a goal of the development program.

The Gasbuggy shot did not yield results which approached the most optimistic expectations, but it did provide some grounds for belief that the technique and technology of nuclear stimulation might be refined to become commercial. The AEC, the Austral Oil Company, and CER Geonuclear Corporation proceeded in their planning of Project Rulison in Colorado. In Gasbuggy, the foundations had been laid. In Rulison, the next logically necessary steps in the development of the technology would be taken. Hopes for the fulfillment of the ideals of the Plowshare Program still ran high.

NOTES

[1]Farmington, New Mexico *Daily Times*, Dec. 31, 1967.

[2]*Ibid.*, Dec. 12, 1967.

[3]Project Gasbuggy Joint Office of Information, *Project Gasbuggy*, Peaceful Nuclear Energy, (PNE-G-4), p. 4, Sept. 15, 1967.

[4]El Paso Natural Gas Co., USAEC, the U.S. Bureau of Mines, and Lawrence Radiation Laboratory, *Project Gasbuggy*, (PNE-G-1), p. 6, El Paso, Texas, May 14, 1965.

[5]*Ibid.*, p. 29, and *Project Gasbuggy*, (PNE-G-4), pp. 2, 3.

[6]*Project Gasbuggy*, (PNE-G-1), pp. 29, 30.

[7]AEC, *Project Gasbuggy Program Information Plan*, Annex 0, (PNE-G-41), Aug. 1967.

[8]*Farmington Daily Times*, Dec. 5, 1967.

[9]*Albuquerque Journal*, Nov. 10, 1967.

[10]*Albuquerque Tribune*, Sept. 28, 1966.

[11]*Farmington Daily Times*, Sept. 23, 1966.

[12]*Ibid.*, Dec. 12, 1967.

[13]*Albuquerque Journal*, Dec. 22, 1966; April 21, 1967.

[14]*Ibid.*, Dec. 3, 1967.

[15]*Albuquerque Tribune*, Dec. 11, 1967.

[16]*Farmington Daily Times*, Dec. 8 and Dec. 10, 1967.

[17]*Albuquerque Journal*, Dec. 11, 1967.

[18]*Ibid.*, Jan. 18, 1968.

[19]*New York Times*, Dec. 11, 1967.

[20]*Farmington Daily Times*, Dec. 12, 1967.

[21]*Albuquerque Tribune*, Dec. 11, 1967.

[22]*Questions and Answers to the Colorado Committee for Environmental Information*, (PNE-G-48), 1969. The Committee asked the AEC what percent of the Gasbuggy device was fusion. The AEC replied that the information was classified but that radionuclide data were available in *Gas Quality Investigation Status Report for Project Gasbuggy*, (UCRL-71314).

[23]L. R. Anspaugh, J. J. Koranda, and W. L. Robison, *Environmental Aspects of Natural Gas Stimulation Experiments with Nuclear Devices*, (PNE-R-53), Aug. 27, 1971, p. 6.

[24]E. Teller, W. K. Talley, G. H. Higgins, and G. W. Johnson, *The Constructive Uses of Nuclear Explosives* (New York: McGraw-Hill, 1968), Chaps. 1, 3, and 6.

[25]K. R. Kase, N. A. Greenhouse, W. J. Silver, and G. R. Norman, *Project Gasbuggy Operational Experiences*, Lawrence Livermore Laboratory Report (UCRL-71356), Jan. 10, 1969.

[26]A. Holzer, *Gasbuggy in Perspective*, Lawrence Livermore Laboratory Report (UCRL-72175), 1970, in *Underground Uses of Nuclear Energy, Pt. 2*, Hearings before the Subcommittee on Air and Water Pollution, Committee on Public Works, U.S. Senate, Aug. 5, 1970, pp. 772-773.

[27]C. H. Atkinson, D. C. Ward, and R. F. Lemon, "Gasbuggy Reservoir Evaluation—1969 Report," in *Proceedings of the ANS Topical Meeting, Engineering with Nuclear Explosives*, Vol. 1, Jan. 1970, pp. 722-731.

[28]Holzer, *Gasbuggy in Perspective*, p. 770.

[29]Atkinson, *Gasbuggy Reservoir Evaluation*, p. 5.

[30]*Ibid.*, p. 5.

[31]*Ibid.*

[32]*Ibid.*

[33]See for example, *Environmental Statement, Rio Blanco Gas Stimulation Project*, USAEC, (WASH-1519), April 1972, pp. 2-6.

[34]Holzer, *Gasbuggy in Perspective*, p. 796.

[35]Teller, *Constructive Uses of Nuclear Energy*, p. 257.

[36]Atkinson, *Gasbuggy Reservoir Evaluation*, p. 4.

[37]Holzer, *Gasbuggy in Perspective*, p. 795.

[38]G. C. Werth, M. D. Nordyke, L. B. Ballou, D. N. Montana, and L. L. Schwartz, *An Analysis of Nuclear Explosive Gas Stimulation and the Program Required for its Development*, Lawrence Livermore Laboratory Report (UCRL-5096), April 20, 1971, p. 30.

[39]Holzer, *Gasbuggy in Perspective*, p. 796.

[40]I. Y. Borg, "Microfacturing in Postshot Gasbuggy Core GB-3," *Nuclear Technology 11*, July 1971, pp. 379-389.

[41]Holzer, *Gasbuggy in Perspective*, p. 791.

[42]*Ibid.*

[43]D. W. Gordon, J. B. Jordan, and B. G. Reagot, *Seismic Analysis of a Nuclear Explosion*, Gasbuggy, (PNE-G-32), Dec. 10, 1967.

[44]S. Glasstone, *Public Safety and Underground Nuclear Detonations*, Technical Information Center, AEC, Oak Ridge, Tenn., (TID-25708), June 1971, p. 111.

[45] Holzer, *Gasbuggy in Perspective*, p. 775.

[46]*Ibid.*, p. 784-787.

[47]F. F. Momyer and C. F. Smith, *Gas Quality Investigation Program Status Report for Project Gasbuggy*, Lawrence Livermore Laboratory Report (UCRL-71374), Sept. 1969, p. 2.

[48]Holzer, *Gasbuggy in Perspective*, pp. 775-777.

[49]C. F. Smith, *Project Gasbuggy Gas Quality Analysis and Evaluation Program: Tabulation of Radiochemical and Chemical Analytical Results*, Lawrence Livermore Laboratory Report (UCRL-50635), April 1969, p. 15.

[50]C. F. Smith, *Non-Gaseous Radioisotopes, Project Gasbuggy Chimney Gas*, (PNE-G-30), Jan. 1969, p. 3.

[51]Holzer, *Gasbuggy in Perspective*, p. 787-789.

[52]Yen Wang, M. D., *Handbook of Radioactive Nuclides* (Cleveland: Chemical Rubber Co., 1969), pp. 621-622.

[53]M. J. Kelley, P. S. Rohwer, and D. G. Jacobs, *Second Quarterly Report on the Theoretical Evaluation of Consumer Products from Project Gasbuggy*, Oak Ridge National Laboratory Report (ORNL-TM-2513), March 1969, p. 2.

[54]*Ibid.*

[55]*Ibid.*

[56]C. J. Barton, D. G. Jacobs, M. J. Kelley, and E. G. Struxness, "Radiological Considerations in the Use of Natural Gas from Nuclearly Stimulated Wells," *Nuclear Technology 11*, July 1971, pp. 335-344.

[57]A. Holzer, "Gasbuggy Experiment," in *Education for Peaceful Uses of Nuclear Explosives*, ed. L. E. Weaver (Tucson: University of Arizona Press, 1970), pp. 23-44.

[58]*Annual Report to Congress of the Atomic Energy Commission for 1970* (Washington, D.C.: U.S. Govt. Printing Office, Jan. 1971), p. 196.

[59]Holzer, *Gasbuggy in Perspective*, p. 795.

[60]Teller, *The Constructive Uses of Nuclear Energy*, p. 3.

[61]*Annual Report of the AEC for 1970*, p. 196.

[62]H. F. Coffer and G. H. Higgins, "Future Contained Nuclear Explosives Experiments," in *Education for Peaceful Uses of Nuclear Explosives*, ed. L. E. Weaver (Tucson: University of Arizona Press, 1970), pp. 45-63.

[63]See for example, *Annual Report to Congress of the Atomic Energy Commission for 1973*, Vol. 1, p. 21, U. S. Govt. Printing Office, Jan. 1974.

[64]Holzer, "Gasbuggy Experiment," p. 44.

[65]USAEC, *Rio Blanco Environmental Statement*, pp. 2-6.

71

Chapter
4
Project Rulison

Project Gasbuggy was one of New Mexico's top news stories in 1967, but in 1969, its Colorado counterpart, Project Rulison, was considered to be even more newsworthy. Newspaper editors throughout Colorado voted Project Rulison the number one news story of the year carried on the United Press International wire service in the state. As the editors noted, the explosion of "a nuclear device more powerful than 40,000 tons of TNT which gouged a cavern deep beneath the Colorado Rockies [was] the most controversial exploration project ever undertaken in the state."[1] No less newsworthy than the ground tremors caused by the blast were the tremors which ran through the body political both before and after detonation.

THE BEGINNINGS

Plans for Project Rulison began to take shape well before the Gasbuggy experiment took place in 1967. In 1965, the Austral Oil Company of Houston, Texas, became interested in the use of nuclear explosives to stimulate gas production from tight or highly compacted geological formations. Austral Oil is a small company which owned several undeveloped oil and natural gas leases, one of which is located near Rifle, Colorado, in the Rulison field.

The Rulison field lies partly in the southern and central sections of Garfield County and to a lesser extent in northeast Mesa County. Figure 4-1 shows the Project Rulison site and nearby communities. The site chosen for detonation is on privately owned land, lying just outside the Battlement Mesa section of the White River National Forest. The closest town, Grand Valley,

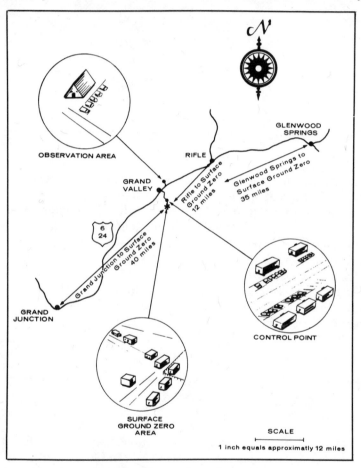

Figure 4-1. Project Rulison area map. (*Project Rulison*, Project Rulison Joint Office of Information, May 1, 1969.)

74

Colorado, is about 6 miles to the northwest and has a population of about 245 people. The nearest more densely populated community is Rifle, about 12 air miles from the site and with a population of about 2,200. Glenwood Springs and Grand Junction are larger population centers, 35 miles to the northeast and 40 miles to the southwest, respectively.

The Mesaverde Formation in the Rulison field is particularly attractive for testing the possibilities of recovery of natural gas by nuclear stimulation technology. Conventional techniques which had been tried had proved incapable of producing gas from the formation at a reasonable rate of return. In 1965, Austral Oil Company acquired additional leaseholds in the Rulison field, bringing its total holdings there to about 36,000 acres. The company then asked the CER Geonuclear Corporation of Las Vegas, Nevada, a company specializing in industrial applications of nuclear science, to prepare a detailed study on the feasibility of using nuclear stimulation.

By early 1966, two test wells had been drilled, and a series of tests were conducted to determine more precisely the gas-producing potential of the Mesaverde reservoir. An estimate of the probable quantity of potentially retrievable natural gas in place indicated that the leaseholds contained sufficient quality and quantity of gas for recovery by nuclear stimulation. In July 1966, a formal letter of intent and application and the Project Rulison feasibility study were submitted to the AEC's Division of Applied Technology. The feasibility study suggested that two 50-kiloton nuclear devices be fired simultaneously underground to produce a chimney 1,600 to 1,700 feet in height and 600 to 800 feet in diameter. If successful, the initial Project Rulison test well could be followed by full field commercial development. A tentative proposal for nuclear stimulation of the entire Rulison field suggested subsequent spacing of wells over every 320 acres which would require more than 100 nuclear detonations of approximately 200 kilotons each.[2]

The brief summary of the *Feasibility Study* did not attempt to justify the use of nuclear explosives for gas recovery in a comprehensive way, though it did note that the United States Government has

> a large vested interest in this project since it owns most of the land and would gain substantial royalties from its production. Thus the government will be interested not

75

only in the development of the nuclear stimulation technique, but in the income derived from gas production. Predicted royalties to the Federal Government alone would more than repay all the money spent on the Plowshare Program since its inception.[3]

In November 1966, Kenneth Brill and H. F. Coffer of CER Geonuclear met with Governor Love of Colorado to discuss the plans that were underway. The following March, a meeting of approximately 700 people was held to provide a general briefing on the nature and purpose of the AEC's Plowshare Program and its proposed current implementation in Project Rulison. Governor John Love chaired the meeting, which included a number of people who might later find themselves, their offices, or agencies involved in the project. Representatives of Western Slope communities, industry, and government were among those invited to attend. Throughout the remainder of 1967, similar briefings and public meetings were held at various locations throughout Colorado.

While these explanatory Plowshare Program meetings were being held, preliminary technical planning also was being carried out. Personnel from the Lawrence Livermore Laboratory inspected the proposed test site with technical support personnel from CER Geonuclear Corporation and the Austral Oil Company. Preliminary specifications for the Project Rulison exploratory well were drawn up, and preparations moved forward in the first six months of 1967 with little indication of either public protest or technical trouble ahead.

During the summer of 1967, liaison work with the appropriate state agencies continued in Colorado. Support and assurances of assistance were sought from the Colorado Department of Public Health and the Department of Natural Resources. The Public Health Department was asked to evaluate the hydrology, or water resources, of the test site with company engineers and scientists in order to preclude the possibility of any adverse effect on the local water supply. The Colorado Department of Natural Resources was consulted on all plans and proposed test procedures which would affect the state's natural gas resources.

On the national level, permits from the U.S. Geological Survey were applied for and obtained. These permits included a Unit Development Agreement and a Unit Operations Agreement, which are required by the Geological Survey of the Department of the Interior before any natural resources may be extracted from federal lands. A Unit Development Agreement legally outlines the land areas that are to be used in a specific project, and a Unit Operation Agreement is a contract specifying the nature of any project that will be undertaken and the methods that will be used. These agreements covered the use of 50,000 acres of approximately one-half public and one-half private lands; they were formally approved on October 4, 1967. The experimental nature of Project Rulison and the time period necessary to complete all work were explicitly recognized in both permits. A termination date provided that if detonation of the nuclear explosive had not taken place by October 4, 1970, both contractual agreements would be void.

On the state level, in January 1968, Austral applied for and received approval of the two agreements by the Colorado Oil and Gas Commission. Field rules for the Project Rulison operations were established, but with the qualification that the applicants' planned use of nuclear devices was contingent upon the authorization of the AEC. Applications for the drilling of wells and for the related ordinary field or site operations were also made to state agencies during this period, as is standard operating procedure in any natural gas exploration project. The usual procedural order of state agency review and approval followed by federal sanction was reversed in this case, perhaps creating a new procedural precedent for projects involving underground nuclear technology. In any case, the ultimate approval of the Unit Operations and Development Agreements by the Oil and Gas Commission were the only formal agreements required by the State of Colorado in advance of the Rulison detonation.

The momentum of the Project was meanwhile accelerated as the Los Alamos Scientific Laboratory [LASL] of New Mexico was assigned by the AEC the task of designing the nuclear device to be used in the project. LASL provided the operations director, scientific advisors and staff necessary to coordinate, direct and carry out the emplacement and firing of the explosive. All LASL personnel worked directly under the Director of Nuclear Operations of the Atomic Energy Commission's Nevada Operations Office. In May 1968, the exploratory well at the Project Rulison site was completed, and the AEC gave formal authorization to the Nevada Operations Office to cooperate with Austral and CER Geonuclear in the preparation of a "Project Definition Report." Preliminary design planning for the project—the nuclear device—had progressed by early fall of 1968

77

to a point where Austral awarded its own subcontract to Superior Drilling Company for drilling the Rulison emplacement hole, about 300 feet north of the original exploratory well. Superior Drilling Company is an affiliate of Signal Drilling Company, Denver, Colorado, which had drilled the exploratory well. Both companies thus had already gained practical, on-the-job experience at the test site. It was necessary to begin drilling operations even before the AEC had signed a contract providing the nuclear device because of the time required to drill a well to a depth of 8,430 feet. However, each foot drilled added to the forward progress of the project, as one step after another took place with little questioning of the ultimate result.[4]

In December, 1968, the Nevada Operations Office and the industrial sponsors finished the *Project Definition Report*. The final document was presented to the Nevada Operations Office for governmental review. Official approval took several months, and in the interval the industrial sponsors maintained their efforts to assure active support and cooperation from Colorado's state agencies and officials. In December, 1968, meetings were held with the Director of the Air and Radiation Hygiene Division of the Colorado Department of Public Health, the Chief Engineer from the Colorado Division of Highways, and with members from the staff of the Governor's Office. The meeting with the Colorado Public Health Department's Director of Air and Radiation Hygiene continued previous liaison efforts, because Public Health Department personnel had toured the Project Rulison site as early as July 1968 to help formulate plans for safety operations. In February 1968, H. Aronson of CER Geonuclear Corporation, a member of the Nevada Operations Office, and Tom Ten Eyck, the State Director of Natural Resources, met to discuss the status of project plans. Other such informational meetings occurred with Roger Hansen, the Executive Director of the Rocky Mountain Center on the Environment, and Dr. Robert Weiner, chairman of the Chemistry Department at the University of Denver. David Miller, a Nevada Operations Office public information officer and Clyde Hampton, the legal counsel in Denver for the Continental Oil Company, were often present at meetings with state officials and other interested parties.

A final and formal contract providing the nuclear device and authorizing joint governmental-industry sponsorship was signed by the AEC on March 26, 1969. Not surprisingly, the contract named Austral Oil Company as the program sponsor for Project Rulison and the CER Geonuclear Corporation as the program manager.[5] On that same day, the director of the Colorado Department of Natural Resources was informed by Austral that the formal contract had been signed and that detonation of the Rulison nuclear device was scheduled for May 22, 1969. There was no specific contract signed for Project Rulison by the Governor, on behalf of the people of Colorado, with the Atomic Energy Commission, the Department of the Interior, or either of the two industrial sponsors of the Program, nor was any such contract required. There was neither then, nor is there now, any firmly established legal basis on which the governor can exercise state authority over nuclear explosions within the state. The 1974 referendum passed by the voters of Colorado requiring popular approval before any nuclear blast can be detonated may provide such a legal basis. But, on March 26, 1969, when the Rulison contract was formally signed by the AEC, plans for the nuclear shot had moved inexorably forward from drawing board to actual implementation.

As in Project Gasbuggy, a Joint Office of Information was established by the AEC which in this instance included the U.S. Bureau of Mines, Austral Oil Company and CER Geonuclear Corporation. Various kinds of informational materials, both in print and film, were produced and made available to inform the people and gain their support for Project Rulison in particular and for the technology of nuclear stimulation in general. As the scheduled date for detonation approached, more and more information was released, especially in the form of news articles and pamphlets. With detonation set for late May 1969, the Joint Office of Information issued a three-part press release in April. It explained the desirability, organization, objectives, and anticipated results of the project. The first of these articles, appearing in the *Rifle Telegram* as well as in other newspapers in April, summarized preliminary planning and activity undertaken to date by the project's sponsors. The second article, printed a week later, was titled "Project Rulison: Objectives and Experimental Details." The basic objective of the project, in the words of the news release, was to "gather data on the potential for commercial development of the Rulison field by application of nuclear stimulation techniques. Since the Rulison field is typical of other gas fields, the results of the experiment would have

79

important bearing on the possibilities for nuclear stimulation in other areas."[6] Specific objectives of Project Rulison included data collection on the following items:

1. pre-shot and post-shot gas production characteristics of the reservoir;

2. the amount and nature of any radioactivity produced in the gas through nuclear stimulation, and the changes in gas composition over a period of time at varying production rates;

3. the effects, if any, of ground motion on man-made structures in the site area; and

4. the cost of nuclear stimulation projects.[7]

The economic data the sponsors hoped to obtain included specific information about drilling and construction costs and the economic requirements for logistic support, as well as the cost of all necessary safety programs. Data of this nature was expected to provide a yardstick for estimating the costs of any eventual full-scale development of the Rulison field. The experimental objectives of Rulison went one step beyond those of Project Gasbuggy, which did not aim to provide economic "yardstick" data. In most technical respects, Rulison objectives differed little from those of the first nuclear stimulation experiment. The technological uncertainties of using nuclear explosives have slowly been reduced over the years, but in a nuclear stimulation project for natural gas recovery, there is so much dependence on the specific conditions at the actual site that each separate project must again consider the entire range of experimental questions. The situation in some respects is not unlike that of drilling for water or oil, or mining for any geological resource. Each well has its own particular set of actual geological conditions. Yet, in other respects, nuclear stimulation projects pose unique problems and potential hazards. How many experiments which create radioactive residue and unique man-made seismic shock are enough to justify the use of the technology? This is a question which was not comprehensively dealt with by either the project sponsors or agencies of the State of Colorado.

If there is a need for a series of tests to refine the technology of nuclear stimulation, there is apparently an equal need for further tests to obtain precise economic information. The economic yardstick accountability of nuclear stimulation experiments for private industry and the integral problem of cost analysis remains obscure even after the completion of Gasbuggy and Rulison, because the AEC has funded initial research and development costs for all Plowshare Projects. Governmental research and development costs were not included in financial reports or analysis until Project Rulison was actually in the developmental and operational stages, when AEC personnel, support crews, and the explosives were needed for the completion of the project. It was therefore stated by the Joint Office of Information that "the funding of Rulison at December 31, 1968, had been solely at non-governmental expense."[8] Further costs of the project after 1968, and after detonation, as given by Austral Oil Company, are listed below but do not include the AEC share of the costs.

1969—Presumed at end of year	$ 6.5 million
November 1970	$ 7.5 million
Flaring and testing underway after detonation of the explosive	
May 1971—Project field operations completed	$11.5 million

By July 15, 1973, almost four years after the Project Rulison explosion occurred, no complete cost breakdown had been available publicly.[9] The given figures, of course, do not reflect any related or prorated costs of outside agencies—for example, those expended by governmental agencies such as the Colorado State Highway Patrol, the State Public Health Service, the Colorado State courts, or even the costs born by witnesses at the hearings. The preceding totals thus reflect only the immediate costs of Project Rulison planning and site development that were borne solely by the Austral Oil Company as the project sponsor. It is uncertain whether CER Geonuclear Corporation costs also are included in these totals, or whether, as program manager, CER also had indirect and nonreimbursed costs distinct from those specified in their contract and reports.

Austral Oil Company received some help in the Rulison undertaking through an agreement made with the Colorado Interstate Gas Company. Colorado Interstate Gas agreed to pay $2 million to Austral and 50 percent of the next four million dollars spent by Austral for additional drilling, testing, and maintenance. In return, Colorado Interstate Gas obtained the right to purchase all gas produced in formations down to the base of the Mesaverde, or about 41,000 acres controlled by Austral in the unit, whether produced by nuclear or conventional means. Also obtained were first refusal rights on gas produced below the Mesaverde, and exclusive option to acquire gas purchase rights on Austral's interest in some 125,000 leasehold acres near Pinedale, Wyoming, should Austral undertake a nuclear stimulation program there within ten years, and certain other rights.[10]

Final plans for Project Rulison called for a fully contained 40-kt nuclear explosive, to be detonated 8,430 feet below the surface in the Mesaverde Formation on the north slope of Battlement Mesa. The planned blast would provide "tremendous energy in a small package that could make nuclear fracturing commercially feasible."[11]

Safety measures undertaken for a blast of this magnitude were dealt with in specific detail in the third major article released by the Joint Information Office. Atomic Energy Commission experience with 270 underground detonations of nuclear explosives, ranging from locations in Mississippi to Nevada to Alaska, were cited as providing both technical information and particular experience in underground blasting. Thus when AEC scientists reached the conclusion that Project Rulison could be "conducted in complete safety, with heat and radioactivity all contained deep within the earth," only theoretical arguments based on the possibility of unforeseeable events could be advanced against the proposed safety regulations.[12]

The news release concluded with the observation that detonation would not cause unacceptable risk to people or property in the area of ground zero. If unavoidable damage did occur, the AEC assumed legal responsibility for payment of any claims that might result from blast-related surface effects. All persons in the immediate area and at the well site area at detonation time were to be protected by safety precautions, which included among other things, closing off the test site, evacuating the surrounding area and all mines within a 40- to 50-mile radius, rerouting of air traffic, rescheduling of trains passing through any canyon or known rock slide areas, and monitoring of the prevailing meteorological conditions to detect any changes in wind flow, whether away from or toward populated areas, in the unlikely event that radioactivity from the detonation should escape into the atmosphere.

In April 1969, the Office of Information also supplied data about the detonation's effects on ground water and the safety of the underground water supply. Studies of analogous shots in Mississippi and New Mexico had shown no harmful effects to the flow or supply of either ground or surface water as a result of underground nuclear detonations. Previous tests and experiments to date indicated that when nuclear explosives were detonated near the depth of the water table under the surface of the ground, ". . . most radioactivity is removed from the water by filtration through the surrounding earth." Or, to quote from the Project Rulison reports, "these experiments show that underground water movement is very slow with a few feet per year. The combination of absorption qualities and slow movement reduces the possibility of radioactivity being carried any significant distance by ground water even if radioactivity should enter the water."[13]

A panel of safety experts worked with the Nevada Operations Office in an advisory capacity in drawing up these reports. In addition, an AEC evaluation panel was appointed which would conduct a comprehensive safety review before approval was given for the actual emplacement of the nuclear device. The U.S. Public Health Service, working with the Colorado Department of Health, was made responsible for monitoring of radiation both before and after detonation. The two health departments were also to assist with evacuation procedures, and with disconnecting existing gas and electrical units in the test-site area. The Environmental Sciences Service Administration's Air Research Laboratory was designated as the responsible agency for gathering and interpreting all weather data. The AEC Project Director would use such data on local weather conditions and climate in making the final decision on the exact time and date of detonation. In addition, a number of other agencies, both public and private, were assigned specific tasks relating to safety, monitoring, and evaluation of test results.

By late April of 1969, there was some possibility that delays in the time schedule of project preparations would force a postponement of the detonation date. Eventually, the schedule was readjusted to an anticipated detonation on September 4, 1969. Between May and September, an unanticipated ground swell of public questions, criticism, and legal opposition arose, so that the future of the project itself hung in the balance.

ACTION AND REACTION, MAY TO SEPTEMBER OF 1969

Rumblings of public doubts about the safety of Project Rulison were heard during its planning in 1968 and 1969, but concerted opposition began to crystallize only when the Colorado Water Pollution Control Commission was actively drawn into the picture. In March 1969, a formal request was made, asking that the Colorado Water Pollution Control Commission "restrain the Rulison Project and all other Plowshare projects from taking place in Colorado until formal hearings are held . . . in order to publicly investigate all aspects of the projects."[14] This request was made by Dr. William A. Colburn, president of the Atomic Storage Corporation, of Denver, to Tom Ten Eyck, Colorado's Director of the Department of Natural Resources and chairman of the Water Pollution Control Commission. Dr. Colburn claimed in his initial request that the Colorado Water Pollution Control Commission had established a precedent earlier for holding such hearings, when an inquiry had been conducted on Atomic Storage Corporation's application to bury chemical and atomic wastes in a deep underground formation near Brush, Colorado. At that hearing, the Atomic Storage Corporation's application was denied on the grounds that such wastes would contaminate underground water supplies. Colburn also noted that his corporation was prepared to show that so far it had not been proven that Project Rulison would be safe, and he asked that a representative of his company be called to testify as a state witness against the project. On the basis of Dr. Colburn's testimony in later lawsuits, it seems that his complaint against Rulison was advanced primarily in the hope of gaining reconsideration for his company's waste disposal plan.[15]

Mr. Tom Ten Eyck forwarded Colburn's request to the other Colorado Water Pollution Control Commission members, and also to the State Attorney General's Office for an opinion on whether the Water Pollution Control Commission did indeed have jurisdiction over the pool of water that would be left by the blast, and therefore did have legal authority to delay the Project Rulison detonation if members of the Commission should decide

84

to hold public hearings. The water pool referred to results from the nuclear explosion which vaporizes water adhering to grains of underground sand. When the steam initially produced by the blast cools and condenses into water, a pool forms at the bottom of the nuclear cavity in the blast chimney.

The Water Pollution Control Commission, in a preliminary hearing on Colburn's request on April 9, postponed a decision on whether to hold public formal hearings on his specific request or on the wider issues of the benefit and safety of Project Rulison. At the preliminary hearing, CER Geonuclear Corporation, as program manager, argued that the Water Pollution Control Commission lacked specific jurisdiction because whatever water was formed by the blasting would not be usable, "beneficial" water, and would not be a part of the general water supply. It was also argued that the water that would be released when the rock was vaporized was negligible, and even the small pool that did remain would be completely contained and separate from other ground waters. Colburn, on the other hand, maintained that blast pressures would force water out of its surroundings, both in the initial effects of the explosive on the surrounding rock, and then after detonation itself, as underground pressures forced the water through the rock fractures. The ultimate destination and possible entry of this water into the entire water distribution system remained an unknown factor, in his opinion.

The arguments presented at this preliminary hearing foreshadowed the kind of conflicting scientific analysis which would recur most frequently in court tests of nuclear stimulation technology and procedure. Both in the court of law and the court of public opinion, decisions had to be made on the basis of conflicting scientific testimony which was essentially beyond ordinary, nontechnical understanding.

After the preliminary meeting, the Water Pollution Control Commission met jointly with the Colorado State Board of Health, and both groups toured the Project Rulison site. At a follow-up meeting on April 24, 1969, the Water Pollution Control Commission made a final decision that it would not hold public hearings on Project Rulison. The Commission was convinced that any contaminated water would not reach usable waters and that therefore hearings were unnecessary. In announcing this decision, Chairman Tom Ten Eyck also observed that perhaps the hazard of "ground shocks which will be created by the explosion are of far more potential concern than pollution of water, since the shock will almost certainly cause some local damage."[16]

During the weeks that the Water Pollution Control Commission was debating the question of whether to hold hearings, the AEC Test Manager from the Nevada Operations Office and representatives of Austral Oil, CER Geonuclear, and the U.S. Bureau of Mines briefed Governor John A. Love on the general status of the project and the results of public meetings that had been held on the project on the Western Slope and in Denver. In several of these public meetings, there was acrimonious debate among scientists as well as those who were not so scientific. Many members of the information team who conducted these meetings indeed had impressive credentials, as did many of those who criticized the project. One of the unfortunate indications of the highly controversial and emotional overtones of nuclear stimulation projects is that they have often involved arguments over the relative expertise of those on different sides of the question.

As increasingly heated debate continued, there was some indecision as to when detonation would actually occur. On May 7, an AEC spokesman said that the May 22 detonation would hold, but on the same day, Tom Ten Eyck told local news media that he had heard from the AEC in Nevada that it would recommend postponement until late summer or early fall.[17]

A possible reason for the delay, which ultimately did occur, may have been concern about potential damage to the Harvey Gap Dam from blast-incurred seismic motions. The dam is an 80-foot-high structure which provides irrigation water for areas in the vicinity of Silt, Colorado. Originally, AEC consultants had recommended lowering the water in the dam to decrease the possibility of damage from seismic motions.[18] But a combination of circumstances may have dictated the need for reconsideration, even though a study of the dam indicated, on the basis of all factors involved, that "it seemed reasonable to conclude that ground motions resulting from the nuclear explosions are unlikely to cause any significant damage...."[19] Mr. Ten Eyck, for example, was quoted as suggesting that it would be best to wait "until the water level would be low in the dam anyway," sometime other than late spring when the danger of rock slides is greatest because of ground frost coming to the surface and creating an unstable rock condition. In addition, he said, the prevailing winds later in the year "would be more likely to blow any radiation that might escape through fissures in the ground away from populated areas."[20]

During May, June, and July, individuals and groups actively opposing detonations began to organize. The aim of such opposition was to make citizens aware of unanswered questions relating to the project and to prevent any nuclear explosion until satisfactory answers were given. The subject of Project Rulison thus became a consistently newsworthy topic. The *Denver Post*, for example, carried an editorial on July 17, 1969, in which the writer approved the intent of the Colorado Department of Public Health to ask a lot more questions before formal approval was granted for the detonation of the nuclear device.[21] Some residents of the Western Slope opposed the Rulison test both in broad concept and specific implementation. A spokesman for this opposition was Mr. Frank Cooley, a lawyer from Meeker who was also a member of Governor Love's Oil Shale Commission. He expressed the concern of some residents living close to the test area, mentioning not only the fear of ground-water contamination, but also the possibility that if the natural gas produced was not of sufficient purity to put into pipelines for the Pacific Northwest area, the gas might be used in thermal plants in the immediate Rulison area to generate electricity.

Local use, he pointed out, might put additional radionuclides into circulation in the air basins and atmospheric systems of western Colorado.[22] The equally undesirable release of radioactivity through flaring of the Rulison gas was again cited by Western Slope opponents as a compelling reason for not proceeding with the project. Another source of opposition came from coal miners on the Western Slope, who felt that mines in the area might be imperiled by blast-created seismic shock waves. An AEC spokesman, at a July informational meeting where this question came up, attempted to allay fears by pointing out that all mines within a 43-mile radius of the surface ground zero point were to be evacuated and that the safety program from the beginning had included extensive mine precautions.

Opposition to the Project on the eastern slope of the Continental Divide in Colorado also began to take shape during the spring. In early May, 20 scientists and professors petitioned the Colorado State Legislature to take action in opposition to Project Rulison. They stated their belief that unanswered questions about the project warranted "a moratorium on this or any other nuclear explosion in the state until such time as the reassurances of the Atomic Energy Commission can be supported by factual evidence and objective consideration by independent analysis."[23] This original group of twenty was known as the Colorado Committee for Environmental Information (CCEI), and was the nucleus of an organization which was to become increasingly active in opposition to Project Rulison. In mid-May, CCEI joined with the Colorado American Civil Liberties Union to present a public panel discussion on Project Rulison. The discussion was held in Boulder, Colorado, a community 20 miles to the northwest of Denver. The University of Colorado and numerous scientific institutions are located in Boulder. The general tenor of the meeting indicated that scientists and lawyers who were present were disturbed by a number of elements in project plans. Several of those present were quoted as expressing regret that the Colorado Water Pollution Control Commission had twice refused to hold public hearings on the matter of detonation of a nuclear explosive within the state.[24]

The student body at the main campus of the University of Colorado in Boulder set up an Anti-Pollution Committee within the student government structure. During the 1969 spring and summer semesters, this committee, working with CCEI, presented a series of articles about Project Rulison in the student newspaper, the *Colorado Daily*. Reprints of the articles were sent to 350 local area families in the Rulison area. A number of other persons also requested, and received, additional information from the student group. Some student representatives made several trips to the Western Slope communities as well, and they assisted in setting up public meetings with speakers from CCEI in order to provide sources of information alternative to those provided by project sponsors.

Opposition groups gained force through organization and through the specific attempt to actively influence public opinion during July and August of 1969. A demand for more information was the main rallying cry, and came principally from three men: Dr. H. Peter Metzger, then a research scientist with Ball Brothers Research Corporation in Boulder, and now a science writer with the *Rocky Mountain News* as well as author of a book on atomic energy; Dr. Robert H. Williams, who was then a physics professor at the University of Colorado; and Dr. Edward Martell, a radio chemist specializing in atmospheric radiation studies at the National Center for Atmospheric Research. These three scientists, who were the original founders of CCEI, said in a press release that they wanted to see all of the facts and analyses of the Rulison plans presented to the public so that Colorado residents "could judge whether the risk of serious long-term environmental contamination should be taken."[25] Among other matters, they particularly stressed the long-term hazards associated with the flaring process. This view was illustrated by citing the experience of the Gasbuggy detonation, which indicated that radioactivity of atmospheric water vapor, at relatively short distances from the New Mexico test site, had doubled during flaring operations. This situation, as the three scientists saw it, adversely affected both plants and grazing animals through distribution of radioactive nuclides in the ecological food-chain cycle.

A further objection was that the explosive to be used in Project Rulison was not a typical Plowshare device, but "an all-fission explosive, using Uranium-235. . . . it [was] expected to produce a much larger quantity of gaseous Krypton-85, about 25% of the gaseous tritium [produced by] Gasbuggy, and large amounts of solid fission by-products which [were expected to] remain in the radioactive melt left underground."[26] CCEI submitted a number of questions to the AEC on these and other matters in mid-July in the hope of quickly obtaining more specific information.

In general, informational material distributed by students and by CCEI both before and after the Rulison detonation, cited familiar concerns about the effects of the detonation on such matters as water pollution, seismic shock, and radiological and general health effects from subsequent flaring. Other considerations were also brought out in this literature which touched upon wider social, legal, and political issues:

> The AEC and the oil companies have repeatedly emphasized that Rulison is an experiment, and claim that is why they have been unable to give answers to questions asked by scientists, concerned citizens and government officials. However, THE PEOPLE OF COLORADO HAVE NEVER BEEN GIVEN THE

OPPORTUNITY TO CONSENT OR REFUSE TO
BE EXPERIMENTED UPON.

Nor in many cases have they been given access to
information that would enable intelligent decisions. . . .
The AEC is acting under Section 31 of the 1954 Atomic
Energy Act. This is a research and development section
and does not provide for commercial application. . . .
The question arises as to whether the federal agencies
are overriding their authority. . . .[27]

Pressure on state agencies and leaders began to build up in the
face of organized opposition and mounting criticism. However,
Governor Love, after a meeting with project sponsors on August
11, 1969, found "no reason to object on grounds of safety."[28] In
addition, the Governor stated that he was not sure whether he had
the power, as Governor, to halt plans for the experiment, since
most of the land in the Rulison site was under federal jurisdiction.
He did concede, however, under questioning, that the state
government does have police power in matters concerning the
health and safety of the people. He added that he suspected the
project would be abandoned if the state made strong protests, but
at this time he could see no reason for such protest.

Meanwhile, new groups joined CCEI in attempting to muster the
force of public opinion to persuade the Governor to use his
authority and influence to secure at least a postponement of the
Rulison explosion. Members of People United to Reclaim the
Environment (PURE) of Boulder, and a Denver-based group,
Citizens Concerned about Rulison, began their own campaigns,
urging people to write, telegraph, or phone the Governor's office
in expression of their opposition. PURE also attempted to
organize a jointly sponsored meeting with environmental
advocates and members of the Joint Office of Information, but
were unsuccessful in this because of previous commitments
undertaken by the sponsors. At the end of one public meeting
held by groups critical of the project, it was suggested that a civil
disobedience sit-in at the test site was perhaps the only means of
securing postponement. Civil disobedience was ultimately
carried out by only a few individuals, but it was indicative of the
frustration of some citizens who were groping for a means of
directly influencing the Rulison decision-making process.

THE COURTS, THE GOVERNOR, AND THE PEOPLE

On August 18, 1969, as the September 4 detonation date drew nearer, attorneys for the American Civil Liberties Union and a group of private citizens filed suit under federal law in Denver District Court. Both temporary and permanent injunctions against detonation were sought. This legal action at the outset was somewhat tardy, for on August 20, 1969, the nuclear devices were lowered into the well, and the well hole itself was sealed. All the individuals who were plaintiffs in the suit were persons who owned property near the site. They included, among others, Richard Crowther, a Denver architect, William Eames, as a private citizen and as the parent of James Smith III. The suit was a class-action procedure brought on behalf of the plaintiffs and all other persons similarly situated. Defendants named in the suit were Dr. Glen T. Seaborg, Chairman of the U.S. Atomic Energy Commission at the time, the Austral Oil Company, and CER Geonuclear Corporation. A second suit was brought by Martin G. Dumont, District Attorney for the Ninth Judicial District of Colorado, which encompasses the Project Rulison site. Defendants in this action were again Chairman Seaborg, Austral, and CER Geonuclear, as well as Claude Hayward, who owned the land on which surface ground zero was located. And finally, a third legal action aimed at stopping detonation was filed a few days later, by the Colorado Open Space Coordinating Council as plaintiff. The Council is a nonprofit public benefit organization concerned principally with environmental issues and related political action in the state. Defendants in this suit were Austral Oil Company and CER Geonuclear Corporation. This last legal request for a stay or halt to the Rulison explosion was brought on behalf of all persons entitled to protection from possible harmful effects resulting from release of artificially created radiation into the biosphere and into natural resources.

On Monday, August 25, in a brief court session just before the hearing on the request for a preliminary injunction postponing the scheduled detonation time and date, Judge William E. Doyle began the proceedings with a reprimand to the plaintiffs' lawyers for the "eleventh-hour" nature of these lawsuits, pointing out that they created a series of problems for the Court and for the defendants' lawyers as well. Judge Doyle observed that since the September 4, 1969, detonation date had been clearly established the preceding July, it was unfortunate that legal actions had been filed so late.[29]

All three cases involved first a request for a preliminary injunction to prevent the Project Rulison blast from taking place. In effect, this is an appeal to the judicial discretion of the Court, or initially to the presiding Judge's discretion and knowledge of the law. The claimant parties suffer under two main burdens of proof which must be sustained from the outset to the Court's satisfaction if a preliminary injunction is to be granted. In these lawsuits, the plaintiffs had to first demonstrate that there was a reasonable probability that they would ultimately be entitled to a permanent injunction preventing detonation of the nuclear explosive. Secondly, they had to show that irreparable damage would occur if the preliminary injunction itself was not granted. Judge Alfred J. Arraj, the trial judge for the hearings on the requests for preliminary injunctions, found that the suing parties did not sustain either burden of proof. The defendants' arguments convinced him that " in all probability" no harm would result from the radioactivity which might be released. Having lost the first round, the plaintiffs immediately appealed the decision.

At this point, political considerations entered the already complex situation. The whole Rulison question became enmeshed in the election campaign of Mark Hogan, who was aspiring to the Democratic nomination for the governorship. Hogan, who at this time was the Lt. Governor, had been elected with and was serving under the Republican incumbent, John Love. If successful in the Democratic primary, he would run against Love in the forthcoming November general election. On August 20, 1969, Mr. Hogan issued a statement saying that he supported the legal efforts to postpone the Project Rulison explosion until "absolute safety" was established. Hogan cited previous "inconclusive results" from Project Gasbuggy, and emphasized the fact that, "Colorado must make it forcefully clear to the Federal government that we do not want this state to be used as an experimental area."[30]

The following day, on August 21, Governor Love told reporters at a press conference that he was "open minded" about possible dangers connected with the Rulison Project, but he added that based on such information as was available to him, there "appeared to be no danger to life or unacceptable property damage which would not be reimbursed." He was "encouraged," he said, by the support he had received for the Project from U.S. Representative Wayne Aspinall, Democrat of Colorado, whose district included the blast site.[31] Representative Aspinall at the time was Chairman of the House Committee on the Interior, which has control over the use of federal land in the country. The Governor further noted that he would have a meeting in the near future with Dr. P. Metzger and members of the CCEI, but he did not expect to receive any information that would cause him to change his position.

At his meeting with both state officials and representatives from CCEI, Governor Love conceded that he was unsure of his position on Project Rulison. Although he had not received any information that would cause him to reverse his previous judgments, there were still some unsettled matters. He specifically wanted reassurance that: (1) no radioactive gases or dusts would be vented into the environment from the well shaft; (2) no harmful side effects would be produced from detonation of the explosives and methods used to remove by-products in the gas; and (3) that small amounts of tritium would not be concentrated in the natural food chain from plants to animals to humans through food consumption. The Governor indicated that if satisfactory answers were not received, it might be wise to postpone the blast, but he repeated that he did not know if he had the legal authority to demand a delay. He expressed confidence, however, that officials in charge of the experiment would abide by his position whether they were legally required to do so or not.

Environmental groups that had been agitating for an open meeting with project sponsors were mollified when a panel discussion with AEC personnel, other governmental agency speakers, and independent experts was scheduled in the auditorium of the Denver Public Library. Speakers urging a delay in the project included Dr. Metzger and Professor Williams from CCEI and Dr. Ernest Sternglass, a radiation physicist from the University of Pittsburgh. Dr. John Emerson, senior industrial hygienist with the Colorado Department of Health, Dr. John Rold, state geologist, and Dr. P. W. Jacoe, chief of the Division of Air, Occupational Health and Industrial Hygiene of the Colorado Health Department, were also present and spoke in their official capacity. The state officials were questioned repeatedly by the environmental and academic representatives,

93

who seemed to feel that these officials, though claiming objectivity, were implicitly endorsing detonation. Dr. Metzger observed that their concern should be "for the public safety, not for the shortage of gas."[32] The meeting in general seemed to provide more heat than light, since both proponents and opponents were convinced of the essential correctness of their respective positions.

In the opinion of CCEI members, the attempt to shed more light on the whole picture was not appreciably influenced by the AEC's "answers to their 29 questions." These answers were formally filed at the Denver Federal Center on August 31, six weeks after they had initially been requested and less than a week before the scheduled detonation. Dr. Metzger announced that his group found the responses to be generally unacceptable and that some of their questions had not been answered at all. The AEC, for its part, stated that detailed information on some matters, such as the design of the nuclear device that was to be used, could not be released to the public because it was classified information.

In the last three days before the scheduled detonation, the appeal of the Rulison suits was heard by the U.S. 10th Circuit Court. The Appeals Court did not reconsider the merits of either side's arguments in depth; it dealt only with the original arguments and briefs to the extent necessary to determine whether there was a reasonable legal basis for overruling the earlier decision. The Appeals Court decision was handed down twelve days after the case was first filed in Denver District Court. Judge Richard Hall, in giving the decision, found that the trial court findings of fact were clear and comprehensive and well supported by the evidence presented. In respect to matters of fact previously presented, the Court found that no errors of law or abuse of judicial discretion had occurred in the earlier findings of the District Court and that there were no grounds on which to base injunctive relief to prevent the detonation.

In essence, the environmental groups had failed to present evidence to the Court that any irreparable environmental damage would occur as a result of the Rulison experiment.[33] The evidence presented by the AEC and its industrial partners was found to be legally satisfactory and to have demonstrated that all possible reasonable safety precautions had been taken. The environmental groups, in other words, had failed to sustain the

94

two major burdens of proof placed upon them. They did not show to the Court's satisfaction that there was a probable right to a future permanent injunction, nor did they show a probable danger to themselves and the class they represented.

The plaintiffs, however, were not deterred by the verdict and filed another suit in Denver U.S. District Court in an attempt to prevent post-detonation re-entry into the well and subsequent flaring of the gas for production testing. Not unexpectedly, this suit was also unsuccessful.

It is difficult to believe that the suing parties had real expectations of obtaining an injunction to prevent the Rulison detonation, especially in view of the late date at which the original suits were filed. They may well have known at the outset that they would probably lose in the courtroom. However, in bringing the entire matter of underground nuclear detonations to the courts, basic issues were raised and some clarification of legal viewpoints was achieved.

Other groups critical of the project took a different tack in opposing detonation. The University of Colorado student Anti-Pollution Committee and the "grass roots" Citizens Concerned About Rulison organization concentrated on influencing public opinion through distribution of literature and public debate of the issues. These groups saw little chance of their making any changes in the project by arguing technical issues with the AEC. Instead, they attacked what they considered the AEC's weak credibility in its handling of public information on radiation resulting from atmospheric weapons testing. The dangers of radiation and some of the AEC's past mistakes in dealing with the production and use of nuclear materials were stressed in an attempt to mobilize public opinion against the project and to persuade the Governor to undertake delaying action. Distributing leaflets, picketing, and organizing small demonstrations were the methods chosen to further bolster opposition forces.

In an extension of this effort, a number of letters and telegrams were dispatched to the Governor's office and to local newspapers expressing opposition and requests for delay. The *Rifle Telegram*, which had become a focal point of Western Slope opposition, again editorially expressed dissatisfaction with the project, asking, "Do we, who live in this area, have a voice in this government-industrial venture???"[34]

In the end, neither the courts, the State of Colorado through its governor, nor the voice of the people secured postponement of the nuclear detonation. Rather, the elements—in particular, an adverse direction of the prevailing winds—caused postponement. A new date was set for 3:00 p.m. on September 10, 1969. On that date, in spite of a last-ditch effort by a small band of protestors to prevent detonation through acts of civil disobedience at the test site, the Project Rulison nuclear device was exploded.

In a somewhat anticlimatic statement in view of the sheer physical force and magnitude of the detonation, which measured 5.5 on the Richter Scale, the AEC announced that the blast had taken place as planned and there was "little unexpected property damage and there was no leakage of radioactive material."[35] Natural gas and electrical services that had been disconnected were restored; rail, air, and highway traffic that had been rerouted or halted was returned to normal; local residents who had been evacuated were advised that it was safe to return to their homes. In some instances, surface structures and buildings suffered slight damage. Dust clouds created by landslides caused by the detonation drifted away to the southeast of the blast test site. The explosion of the Project Rulison nuclear device 8,430 feet below the Western Slope surface of the Rocky Mountains was now history. But, unlike the dust resettling above the blasting area, the criticism which had repeatedly plagued sponsors of Project Rulison did not blow away. Technical evaluations of the results of the project could not be undertaken until the natural gas released from its underground rock formations by the nuclear explosive had been flared. And this process became the subject of new litigation.

By early October of 1969, Judge Arraj was once again the presiding judge for District Court litigation on the subject of Project Rulison. Despite the earlier denial of a temporary injunction to halt detonation, the original plaintiffs pressed forward with new litigation in an attempt to secure a permanent injunction to prevent the scheduled re-entry into the Rulison well. As planned by the sponsors, re-entry was to be achieved through another well, known as the R-EX well, situated about 285 feet southwest of the Rulison well. The plaintiffs were also asking that flaring of the Rulison gas through the R-EX well be prohibited.

96

Early in the hearing of the case, Judge Arraj scheduled oral arguments on the defendants' motion in which they asked for a summary judgment in their favor in all three suits pending against them. Plaintiffs and defendants in two of the suits were identical to the parties in the original legal action which sought to prevent detonation. The motion for summary judgment was denied, and hearings and trial on the merits for a permanent injunction forbidding re-entry and flaring were begun by early January of 1970.

Local newspapers covered some elements in the controversy over flaring before the beginning of the trial. The *Denver Post*, for example, reported in November 1969 that Dr. Seaborg, former chairman of the AEC, had acknowledged that "some tritium and krypton-85 would be released into the atmosphere through flaring, but not in harmful amounts."[36] During the course of the trial, the sponsor's view of the basic safety of flaring was reported, as was the view of the plaintiffs, who felt that the safety of the procedure was far from assured. Both public and judge, in short, were once again faced with a bewildering variety of scientific opinion.

Final arguments in this litigation were presented on February 20, 1970. The Court's opinion and memorandum were handed down on March 11, 1970.[37] Though the plaintiffs' request for a permanent injunction was denied, the court's opinion contained a much more complete and articulate exposition of the legal issues involved than did the opinions in the first law suits, which necessarily had to be completed within a very short time after initial filing. The plaintiffs, though rebuffed in their attempt to gain a permanent injunction, felt that they had won some ground which in the future might provide precedent for those who wished to force federal agencies into environmental accountability.

The most significant legal issues raised in this respect concerned: (1) their standing or right to sue a federal agency, (2) the question of whether the AEC was acting within the bounds of Congressional authorization in carrying out Project Rulison, and (3) the scope of judicial review as it applied to federal agencies. Each of these issues raised important questions in regard to the role of environmental public interest groups and respective state and federal prerogatives and responsibilities in this area of concern.

In order to acquire standing before the Court, or for their cases even to be accepted for hearing, the plaintiffs had to show that they had a valid interest in the issues raised which entitled them to legal protection. Then it had to be shown that this interest might be directly threatened by the actions of the defendants. These basically formal requirements insure that the court is presented with a specific controversy between genuinely adverse interests. With relative ease, the judge decided that the plaintiffs and the Colorado Open Space Coordinating Council had established standing. This decision was based on their allegations of interest and the alleged charges they brought against the AEC, Austral, and CER. In the first two suits, the suing parties were granted legal standing under the general equitable jurisdiction of the Court. Although the Colorado Open Space Coordinating Council could have secured formal standing on these general grounds alone, the Court added that they also had a legal right to sue under the specific requirements of the Administrative Procedure Act of 1966.

It was further held that the AEC could not rely on the doctrine of sovereign immunity, or the assumption that a federal agency is immune to legal action brought against it. In Judge Arraj's view, a complaint against a federal agency of government is permissible if the plaintiff's allegations indicate reasonable grounds for believing that the agency acted beyond the scope of its legal authority.[38] In the preliminary opinion, the judge also found that the court had authority to determine whether the AEC had acted within the scope of its statutory powers as defined by Congress.

The plaintiffs felt that these findings were particularly significant, for in the past, courts had sometimes denied standing to public interest groups seeking to challenge the authority of an agency of the federal government. The fact that a large agency, such as the AEC, which claimed a virtual monopoly over the materials and expertise required for its functions, could be confronted in court was gratifying to the plaintiffs, regardless of the ultimate decision.

Specifically, the Court ruled that it could indeed examine the question as to whether the proposed flaring of the Rulison natural gas was planned with due regard for the protection of public health and safety as required by the Atomic Energy Act of 1954, Section 31 (d). This portion of the U.S. Code and the

Atomic Energy Act of 1954 defines the limits of all activities of the U.S. Atomic Energy Commission, setting forth the responsibilities of the agency and delineating broadly the powers that may be legitimately exercised. The defending parties in these suits argued, unsuccessfully, that such questions as these were essentially political ones, and therefore they could not justifiably be heard by the court. In considering this matter and the arguments presented by the defense motions, the Court confined itself solely to considering the question of whether Atomic Energy Commission health and safety arrangements might have constituted a legal abuse of a federal agency's discretionary power as authorized by Congressional statute. Judge Arraj noted in this respect that the AEC is in almost exclusive possession of experience and expertise in areas of health and safety related to nuclear explosives. It is therefore necessary to allow some review of agency actions, whether by members of the general public or by other concerned agencies, in order to make sure there is compliance with the standards set by law. Thus, although the court did not set down a precise demarcation of its powers of judicial review, the preliminary rulings did allow an evironmental group to question federal administrative practice.

In effect, the finding that these post-detonation Rulison cases were at least open to court review, regardless of the final outcome, fell into a long tradition of checks and balances in the history of American government. While such an avenue of outside legal review eventually may produce a decision in the defendant's favor, the preliminary willingness to accept such litigation in his court placed Judge Arraj firmly in accord with the weight of American legal history and practice. Both the plaintiffs and the defendants would now have the opportunity, denied to them in the earlier suits, to present detailed evidence and arguments based on fully prepared legal research. The preliminary findings, as well, indicated clearly to the AEC that it did not have *carte blanche* in the exercise of discretionary power.

In considering the merits of the case, the court extensively examined the factual issues involved in the project plans for protection of public health, safety, and welfare. The arguments hinged on whether it was reasonably sure that there would be no temporary or permanent harm to the public resulting from re-entry of the well and flaring of the gas. The plaintiffs alleged that a credibility gap existed because the AEC had said it would operate in a specific manner but apparently had proceeded to do something quite different. The specific allegation was that the defendants had not adhered to their own safety plans for alternative test procedures in case wind direction and air movement varied from the standards required for public safety. It was asserted that this issue was important in determining if radioactive particles produced by the detonation would fall within or outside of the predetermined optimum safety area. Judge Arraj rejected this argument, which was based on complex mathematical detail, pointing out that the sponsors had reasonably followed indicated safety procedures.

The court then considered whether the defendants were required to explore other feasible alternatives to flaring, before selecting that particular method of disposing of radioactive natural gas and resulting radioactive by-products. Judge Arraj was satisfied that project planners had investigated alternatives to flaring, though it was not specified in the legal testimony at what point in plans or actual operations such alternatives were considered. It was found that flaring had legitimately been selected as the safest and most economically feasible method of disposal of radioactive residues and wastes created by the test.

Judge Arraj also extensively considered arguments of plaintiffs and defendants and the testimony of expert witnesses about the reasonableness of the radiation and safety standards on which Rulison safety plans and procedures were based. Among other points, the plaintiffs unsuccessfully argued that the Federal Radiation Council standards of 1957 required re-examination in the light of new scientific studies, particularly those concerning the correlation between chromosomal aberration and irradiation. The court concluded that it could only require the AEC and the corporate defendants to act on the basis of "the best available knowledge." And, in 1969 and 1970, the best available studies on radiation safety were legally those set forth in the Radiation Council's standards for permissible exposure and dose.

In conclusion, both the permanent and temporary injunctions requested against re-entry of the well and flaring of the gas were denied. The defendants were thus permitted by law to proceed with the Rulison plans. They were, however, ordered to follow the original Project Rulison Definition Plan and not begin flaring operations for six months; this time period would allow most short-lived radionuclides to decay and thus lessen any chance of adverse effects on public health and safety. The court's issuance of specific further orders to the AEC and the corporate sponsors and its explicit retention of jurisdiction over the cases, and thus over project plans and development, are somewhat unusual actions. The defendants in the litigation in effect would continue to be directly accountable for their performance to the court, "at the behest of the plaintiffs."

Perhaps the most significant end result of the Rulison litigation is that agency accountability to an independent judicial institution was established and specifically retained through the hearing of the merits of the arguments by the court. The plaintiffs succeeded in rendering the Atomic Energy Commission, its planning processes and its safety precautions, and those of private companies contracting with a federal agency in such a mutual project, open to direct question and review in court. Project Rulison thus was not halted, but its architects were forced to defend their plans and practices in a public forum where they could be openly questioned and examined. The fact that the Court took the unusual step of retaining jurisdiction over agency activities related to the project meant that its procedures and results could not be obscured or withheld from public scrutiny. This precedent could be of more far reaching significance than the specific decisions elicited by the Rulison cases.

Despite the actual loss of their suits, the plaintiffs won a greater measure of disclosure about the nature of agency processes than was evident at the time the suits were initially filed. In addition, the sponsors were required to give the court an outline of procedures they would use in order to meaningfully inform the general public about project activities. The court was to be provided with all data resulting from project testing and analysis, and this information was to be published as soon as it became available. Finally, in affirming both original and continuing jurisdiction over the defendants and the subject matter of the dispute, some assurance was given that an experimental technology, even in its developmental stages, would be reasonably and safely executed. In short, it was demonstrated that it is legally possible and permissible to try a new technology, as it relates to and affects social welfare, in a court of law.

As the Rulison court test was in progress, opponents of the project continued attempts to prevent flaring. The students of the Anti-Pollution Committee at the University of Colorado continued to distribute informational material on radiation hazards associated with flaring. Some Aspen residents assembled in the local Wheeler Opera House to express their opposition to any flaring, while another group of Western Slope residents, the Delta Citizens Concerned About Radiation, proceeded in a motorcade to Grand Junction in mid-July of 1970 to add their voices to the protest movement. The *Rifle Telegram* again editorially criticized both the AEC and its industrial partners, and this time cited the Rifle City Council as well for its alleged failure to fully consider the potentially adverse effects of the project.

In late April 1970, the City Council of Rifle sent out 250 questionnaires with city water bills to survey local opinion on the subject of the use of nuclear stimulation technology in their area. A majority of the residents (188) responded by returning the survey cards; of these, 59 percent indicated that they favored the "development of our natural resources in this area by nuclear detonation," 29 percent responded unfavorably, 10 percent conditionally favored the project, and 2 percent of the questionnaires were returned blank.[39] This survey, though not comprehensive, was the only attempt made to register public opinion before or after the Rulison detonation. It does, however, provide some indication that the majority of local residents in the Rifle area within a year after the event favored nuclear exploitation of these natural gas resources.

Flaring of the natural gas produced in Project Rulison proceeded at intervals in 1970 and 1971. During this period, field operations centered on testing procedures to collect the technical data necessary for subsequent analysis. The question of whether the Rulison well would ever be reactivated and the gas produced from it would ever be marketed remained a moot point for years after the original detonation.[40]

In the years after the Rulison field operation was put in mothballs, technical data and results have become available. A brief summary of these results is presented in the next section to provide a general understanding of the process. The ultimate success of the technology, however, can be judged only after safe and economically viable use of gas stimulated by nuclear means by consumers in the state and the nation has occurred.

THE RULISON RESULTS

On September 10, 1969, a nuclear device was detonated 8,430 feet underground in a 2,500-feet-thick, gas-bearing Mesaverde Formation at Grand Valley in Colorado. It is estimated that the Rulison field contains 60,000 productive acres with 8 trillion cubic feet of gas in place. Clearly, there is sufficient gas in this field to significantly bolster the nation's diminishing gas supplies if the source can be tapped safely and economically. Figure 4-2 is a schematic cross section of the rock strata at the Rulison area. A portion of the Mesaverde Formation, which underlies a large area of western Colorado, was the object of the Rulison stimulation experiment. To stimulate the gas flow, a nominal 40 kiloton nuclear all-fission device was used. After detonation, it was determined that the actual yield of the device was 43 ± 8 kilotons. The Rulison experiment added technological data and scientific knowledge about the technique of nuclear stimulation but, as shown later, it did not demonstrate the economic feasibility of the technique, and it clouded the future for other such projects.

The data-recording equipment and the operational procedures for Project Rulison were similar to those used in Project Gasbuggy. The explosive canister was smaller than the Gasbuggy device. It measured 15 feet long, 9 inches in diameter, and weighed only 1,200 pounds. The size of a nuclear stimulation device is important, because the larger and heavier it is, the larger and more expensive will be the hole and the canister. For the Rulison device, a 15-inch-diameter hole was drilled, cased to 8,700 feet and then filled with concrete up to the detonation point (8,430 feet). After detonation, a chimney formed in the rock, as expected. Table 4-1 compares the expected and actual chimney dimensions. As in Gasbuggy, no physical measurement of the chimney was made, but its dimensions were estimated from measurements of flow rates and bottom-hole pressures. Thus, even though the explosive power of the device was larger than anticipated, the cavity was somewhat smaller than expected. The AEC's explanation for the variation between the expected and actual chimney dimensions is that the depth of burial had a greater effect than had been anticipated. As in Gasbuggy, predictions of the effect of the blast proved optimistic. Because the chimney size and the extent of fracturing directly determine

103

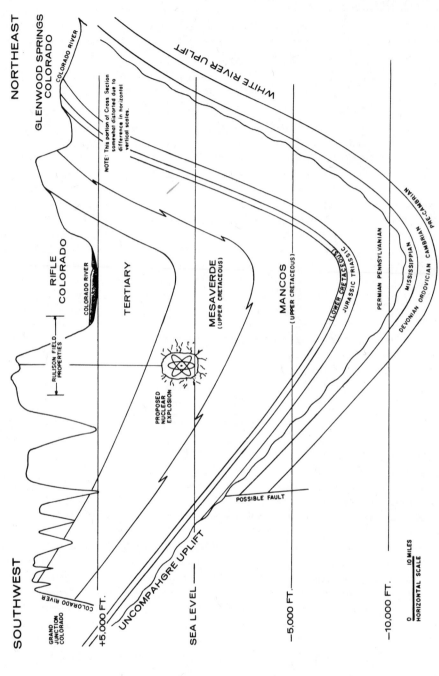

Figure 4-2. Schematic cross section of the Rulison area. (Project Rulison Joint Office of Information, *Project Rulison*, May 1, 1969, p. 5.)

TABLE 4-1
PREDICTED AND ACTUAL CAVITY DIMENSIONS FOR
PROJECT RULISON

	Predicted[a]			Actual
	Max.	Mean	Min.	
Chimney radius, feet	108	90	72	76[b]
Fracturing radius, feet	580	485	390	222-370[c]
Chimney height, feet	451	376	301	350[d]
				250[e]
Chimney volume void, millions of cubic feet	5.28	3.05	1.56	1.80[b]
Total cavity volume, millions of cubic feet	16.7	9.57	4.90	6.35[f]

[a]M. Reynolds, Jr., B. G. Bray, and R. L. Mann, "Project Rulison: A Status Report," Paper No. SPE-3191, presented at Eastern Regional Meeting of the Society of Petroleum Engineers, Pittsburgh, Pa., Nov. 5-6, 1970.

[b]Milo D. Nordyke, "Underground Engineering Applications, Concepts and Experience," Lawrence Livermore Laboratory Report (UCRL-72769), Dec. 15, 1970.

[c]M. Reynolds, Jr., private commun., dated July 31, 1972.

[d]M. Reynolds, Jr., "Project Rulison—Summary of Results and Analyses," *Nuclear Technology*, 14, May 1972, pp. 187-193.

[e]R. W. Taylor, "Thermal Effects of Underground Nuclear Explosions," *Nuclear Technology*, 18, May 1973, pp. 185-193.

[f]Calculated on the basis of a cylinder with a radius of 76 feet and height of 350 feet.

the rate of gas flow, the results appear to have been disappointing. The discrepancies could not have been totally unexpected, however, because the blast occurred in a geologic environment for which little previous work had been done.

The Rulison gas well produced 455 million standard cubic feet of gas in a period of 202 days during which gas was withdrawn in four separate tests totaling 108 days. Of the 455 million standard cubic feet, approximately 50 million cubic feet was water vapor and 120 million feet was carbon dioxide. Therefore, the saleable, or dry, carbon-dioxide-free gas was 285 million cubic feet or 62.6 percent of the total amount of gas produced. Conventional stimulation, production, and distribution of natural gas includes processing to remove carbon dioxide and water, which do not contribute to the heating value of the gas and are not saleable products.

On the basis of the measured gas flow, early predictions were that Rulison could produce 5 to 7 billion cubic feet of gas during a 20-year production lifetime.[41] Later estimates, based upon these same data, were much lower, however. Rubin *et al.* estimated a productivity of 1.8 billion cubic feet, and indicated that this result agreed with a prediction by CER Geonuclear.[42] Estimates were that conventionally stimulated wells in the field would produce in the range one-half to one billion cubic feet of gas in 20 years.[43] The permeability at the Rulison location was .011 millidarcies as compared to the Rulison field average of .050 millidarcies.[44] This indicates that, on the average, stimulation of the rest of the field with the same techniques used in Rulison might yield more gas than the Rulison shot.

Preliminary reports on the results of Rulison indicated that the experiment had demonstrated the technical feasibility of nuclear stimulation of gas in the Rulison field, even though an evaluation of the economic potential of the results had not been made. Five years later, neither Austral nor AEC has published an economic analysis of Project Rulison, although hypothetical projections were made based in part on these results. It seems curious, however, that if Project Rulison had shown that it was economically feasible to produce gas, field development had not been initiated. Five years after the blast, the Rulison gas still remains in the well and no plans have been announced for further development of the field.

As mentioned earlier, Austral was able to obtain some financing by selling gas rights to Colorado Interstate Gas Company. CIG paid Austral $2 million when the Rulison well was successfully reentered and agreed to pay (up to a maximum of $2 million) one-half of the additional expenses incurred in developing and maintaining the Rulison field. In return, CIG was granted the right to purchase the gas Austral produced in that unit as well as from some other Austral leases at 20 cents per thousand cubic feet, or at the prevailing area price, whichever is higher. CIG could then recoup the advances made to Austral, with interest at 7 percent, by means of a credit against 50 percent of the purchase price to which Austral would otherwise be entitled for the gas produced. Since no production has taken place, CIG has in effect given Austral a long-term loan on which Austral makes no payments until gas is produced. CIG certainly would not have advanced that kind of money unless it expected substantial gas production. At present, it looks as if it will be a long time before those expectations are fulfilled.

Although it is obvious that gas production was stimulated at Rulison, to what degree stimulation occurred is uncertain. The AEC did not publicly claim that five to eight times greater production was obtained from nuclear stimulation than from conventional stimulation at Rulison wells in the same field. But judging from the cavity dimensions and from Austral's inaction with regard to full field development, the stimulation appears to have been less effective than anticipated.

The chemical composition of gas produced from the Rulison well was similar to that produced from the Gasbuggy well. Initial hydrocarbon concentrations as well as the heating value of the gas were lower than in gas produced in the field by conventional techniques. Figure 4-3 is a plot of the concentrations of methane and carbon dioxide versus the total gas production. Since methane is the primary hydrocarbon contained in natural gas, it is a good measure of the heating value and quality of the gas. The lower the methane concentration, the lower the gas quality and the lower its economic value. Carbon dioxide is the primary contaminant in natural gas; it does not add to the heating value and must be removed before the gas can be sold. An extrapolation of the curves in Figure 4-3 to determine how long the quality of the gas will be affected by carbon dioxide dilution is speculative. Although several explanations about the high carbon dioxide concentrations have been given,[45] whether the gas from the nuclear Rulison well will ever be of a quality comparable to the rest of the field is still in question.

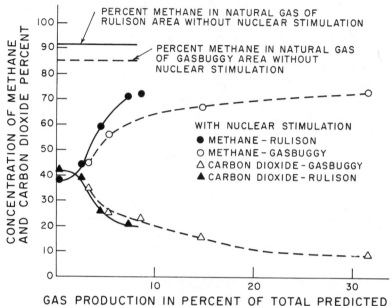

Figure 4-3. Methane and carbon dioxide contents of Gasbuggy and Rulison gases. The concentration of methane is an indication of the heating quality of the gas. The two horizontal lines represent the methane concentrations of the gas that could have been produced from the two wells by conventional techniques. Carbon dioxide is of no value and must be removed before the gas is sold. (Figure prepared by the present authors from data given by M. Reynolds, Jr., "Project Rulison—Summary of Results and Analyses," *Nuclear Technology*, 14: 187-193 (May 1972) and by F. Holzer, *Gasbuggy in Perspective*, Lawrence Livermore Laboratory Report UCRL-72175 (1970).)

Figure 4-4 presents the methane and carbon dioxide concentrations as a function of the percent of total expected production. This graph shows that the quality of the Rulison gas improved more rapidly than did the quality of Gasbuggy gas. While it is not likely that the Gasbuggy well will ever produce gas of the same quality as the gas produced in conventional wells in the Gasbuggy field, the Rulison well might some day produce gas of the same quality as other wells in the Rulison field.

The switch from the fusion device used in Gasbuggy to an all fission device in the Rulison experiment indicates that a reduction in gaseous radioactivity was necessary. While a fission device produces more total radioactivity, it produces less gaseous radioactivity, particularly tritium. A military fission device is often called a "dirty bomb" because it produces more

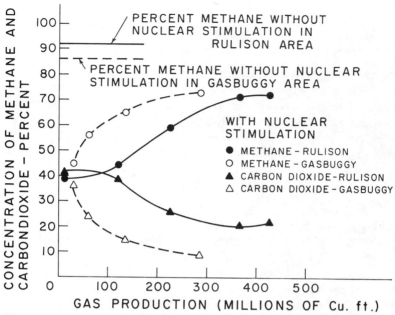

Figure 4-4. Methane and carbon dioxide contents. (Prepared by present authors from the same data as those in Figure 4-3.)

radioactivity than a fusion device of the same size. But since only gaseous radioactivity was expected to reach the surface, a fission device was considered preferable for use in natural gas stimulation.

The United States Public Health Service concluded that contamination of ground water by solid radionuclides deposited in the Rulison nuclear chimney as a result of an underground stimulation blast was highly improbable.[46] It was predicted that the radionuclides of most concern, cesium-137 and strontium-90, as well as tritium, would move only a few feet through the formation rock before decaying to a point well within AEC safety standards. It was claimed that the calculation assumed "worst case" conditions of a water-filled chimney and uniform mixing of all the nuclides deposited in the chimney.

Observations made in 1969 by the U.S. Geological Survey during drilling of the Rulison Experimental Hole suggested that little mobile water existed in the formations penetrated by the hole.[47]

However, the USGS report cautioned that there was insufficient evidence to conclude that there would be no mobile water in the fracture zone surrounding the chimney.

The AEC expected that after the Rulison blast, most of the solid radionuclides would be incorporated into glass-like solidified rock at the bottom of the cavity where they cannot be dissolved and carried along with mobile underground water. Although the Rulison cavity was not actually examined after the blast to verify the expected entrapment of the solid radionuclides, it was assumed to exist because physical examination of cavities at the Nevada Test Site had shown that such entrapment had occurred there.

The AEC's calculation showed that even if a water velocity 200 times larger than calculated under worst case conditions is assumed, the radionuclides can travel only three miles before decaying to a state below AEC guidelines.[48] The nearest source of domestic water to the Rulison Project is three miles from the site, but 6,000 feet above the chimney. Movement of radionuclides from the chimney into this water or into the Colorado River, 5,000 feet above the chimney and five miles away, appeared unlikely. Essentially, then, the AEC concluded that even under worst case conditions the solid radionuclides would remain in place in the cavity until after they had decayed to nonradioactive elements. No circumstances were envisioned which would alter the validity of this conclusion.

Even though solid radionuclides were not expected to escape, the gaseous radionuclides could easily reach the atmosphere during flaring and would be released during any production activities. Flow testing and subsequent flaring of Rulison gas took place from October 4, 1970, until April 23, 1971. During this period of 202 days, 108 days of flaring operations were carried out, and 455 million standard cubic feet of gas were flared in four separate tests. The total radioactivity release in these flarings was 1,064 curies of krypton-85 and 2,824 curies of tritium. These values are 96 percent of the initial krypton-85 total of 1,100 curies produced by the blast and 28 percent of the tritium total of 10,000 curies. The gas also contained some carbon-14, argon-37, argon-39, and mercury-203, but the concentrations of these radionuclides was so low that significant dosages to humans from them was not considered possible.[49]

110

Pre-shot predictions of the upper-limit radiation risks to the local population from flaring were made by W. L. Robison and L. R. Anspaugh.[50] Later work by the same authors indicated that the pre-shot estimates were higher than the post-shot calculations by a factor of 10,000.[51] Both the pre-shot and post-shot estimated dosages are presented in Table 4-2.

TABLE 4-2
PREDICTED RADIATION DOSAGES TO THE LOCAL POPULATION FROM PROJECT RULISON DUE TO THE RELEASE OF TRITIUM

	Pre-Shot* Distance from source (km)	mrem	Post-Shot† No distance specified (mrem)
Air inhalation and subversion	5 15	0.006 0.002	0.00052
Milk	5 15	30. 10.	0.00011
Food	5 15	2. 0.6	0.00181
Total (Whole body)	5 15	32. 10.6	0.0024

* See end note 50.
† See end note 51.

Unfortunately, the methods used by these same authors in the two reports differ so drastically that on the basis of these reports it was not possible to compare the results using the same method of calculation. Table 4-2 presents therefore only the unverified numerical results.

Maybe the most disturbing aspect of the radiation data is that the AEC scientists could differ by a factor of 10,000. While it is certainly the best policy to err on the side of overestimating the dangers—is it in the public interest to go ahead with full field development, which means literally thousands of blasts the size of Rulison, using a technology that still contains so much uncertainty? The AEC concludes, on the basis of Federal Radiation Council Standards and their calculations, that the dosage received from flaring is not significant. But no specific radiation standards were set for Project Rulison, possibly

111

because the issuance of new guidelines for a one-shot experimental project, such as Rulison, would be premature. On the other hand, the AEC has set new emission guidelines for nuclear reactors because such guidelines are necessary for an on-going and increasingly important activity.[52]

For nuclear reactors, the guidelines place an upper limit of 10 mrem per year at the power plant boundary from noble gas emissions and 5 mrem per year as the result of discharge of liquid wastes. These guidelines do not consider tritium dosages because very little tritium is produced by reactors. However, if for purposes of comparison tritium is considered to be a noble gas (that is, in the form of gaseous hydrogen) or a liquid (that is, in the form of tritiated water), the pre-shot Rulison flaring predictions would exceed either of these guidelines. In view of the apparent uncertainty about the actual dosages, the AEC might have to find ways of reducing radiation exposure caused by flaring in order to be consistent with the reactor guidelines, if full field development of the Rulison or any other field is to be undertaken by nuclear stimulation. In addition to the radiation released by flaring, the dosage received from possible use of the gas is an important consideration. If nuclear stimulation of natural gas is to augment our gas supplies, the gas produced must be useable safely.

In calculating hypothetical radiation dosages from use of the Rulison gas, the AEC investigators used a somewhat different model than that which was used in the Gasbuggy calculations. A major assumption of the Gasbuggy calculations was that the gas would be used solely in unvented home heaters. The use of such home heaters is illegal in Colorado. The only unvented appliances that are commonly used in the state are gas ranges and refrigerators. In the Rulison calculations, the investigators determined dosages by calculating atmospheric concentrations of tritium resulting from gas use in residences and commercial establishments and from unvented home ranges and refrigerators. For comparative purposes, the hypothetical radiation dosages that would be received from the use of Gasbuggy gas can be calculated within the framework of the model used in the Rulison analysis. For these comparative calculations, one must assume that Gasbuggy gas is transported to the Rulison site. The model also assumes that the Rulison and Gasbuggy production rates are one million cubic feet per day for a three-year period, that after a three-year period essentially all of

112

the tritium and krypton-85 are removed from the chimney, and that the gas is introduced into the distribution systems of the Rocky Mountain Natural Gas Company and the Western Slope Gas Company.

Dosages are determined by the tritium concentrations in the gas and the amount of gas used in particular geographic areas. Thus, Aspen residents will receive a relatively high dose in comparison to residents of some other locations because of the great consumption of gas in that locale due to lower temperatures. The additional 0.58 mrem received by Aspen residents is small in comparison to the additional dosage that is uniformly received by individuals who live at high altitudes as a result of cosmic radiation. A two-week stay at Aspen will result in an increased dosage of approximately 4 mrem compared to sea level.

Table 4-3 presents the predicted dosages for several locations. The concentration of tritium for a year-one dosage is known because radioactive analysis was made of the flared gas. Concentrations for year-two are somewhat speculative because they are predicted from the trend of past concentrations (that is, the year-one flared gas). Year-two dosages are approximations of the dosages that would be received if the Rulison gas currently in the well is used. Year-one dosages are those that would have been received if the initial Rulison gas had been used for heating with no flaring.

The dosages reported here for use of the gas in homes with entirely vented appliances are small compared to background. In light of the AEC reactor guidelines, the dosages from the gas used in unvented appliances are probably unacceptable for long-term use. If the gas currently in the well is used, the radiation dosages will be within the limits set in the nuclear-reactor-emission guidelines. Commercial use of the first 455 million cubic feet produced by Rulison would have been prohibited had these guidelines been applied, unless the gas was diluted to an even greater degree.

Radiation from the Rulison blast is impossible to feel or see, but the shock wave and resulting seismic damage accompanying it was easy to detect. The predictions which were made for the seismic motion which would result from the Rulison blast were based primarily on Gasbuggy results. The seismic data resulting from the 95 experiments conducted at the Nevada Test Site

TABLE 4-3
RADIATION DOSAGES, IN MREM, FROM THE
HYPOTHETICAL USE OF RADIOACTIVE NATURAL GAS

	Aspen Pop. 2,404	Delta Pop. 3,694	Rifle Pop. 2,150	Grand Valley Pop. 270
Fraction of contaminated gas in the distribution system	0.10	0.09	0.37	0.69
Year-one doses received by the entire population from residential gas use[a]	0.58	0.18	0.66	0.28
Year-two doses received by the entire population from residential gas use[b]	0.12	< 0.01	--	--
Year-one doses from industrial use[a]	--	0.10	0.24	--
Year-one doses from refrigerator and home range[c]	5.7	--	23.	39.

(margin label: RULISON)

Source of Rulison Data: C. G. Barton, R. E. Morse, and S. R. Hanna in the Atomic Energy Commission's *Quarterly Progress Report on Radiological Safety of Peaceful Uses of Nuclear Explosives: Hypo-Exposures to Rulison Gas* (ORNL-TM-3601), October 1971.

[a]Assumes dry CO_2-free gas perfectly mixed with inflowing "clean" gas. The average tritium concentration in Rulison gas is 107 p Ci/cm^3.

[b]Assumes dry CO_2-free gas perfectly mixed with inflowing "clean" gas. The average tritium concentration in Rulison gas is 14 p Ci/cm^3.

[c]Assumes dry CO_2-free gas with an average tritium concentration of 107 p Ci/cm^3, daily consumption of 45.6 ft^3, a home air change rate of one per hour and home occupancy of 70%.

114

(NTS) differed significantly from the Gasbuggy results because of the different geological media involved.[53] The Gasbuggy detonation was the first experience with an underground nuclear explosion in the kind of sedimentary, sand-shale medium characteristic of both the Gasbuggy and Rulison sites.

The observed accelerations in Rulison were somewhat greater close to ground zero and somewhat less farther away than were predicted on the basis of Gasbuggy data. Ground acceleration is important because it can cause damage to surface structures. The primary factors contributing to the inadequacy of predictions appear to be the relatively greater depth of burial of the Rulison device, and the shale-siltstone geologic environment. "Unlike the predictions, the observed [Rulison] ground motions did not attenuate proportionately with distance, in all sections of the project area."[54] The seismic waves were propagated at different rates corresponding to the outline of the Rulison basin. This behavior is probably accounted for by variations in the geological-geophysical character of the rock.

Some representative peak acceleration data are presented in Table 4-4. The difference in accelerations for approximately equidistant locations is appreciable, particularly in vertical components. Most of this variation may be explained by differences in elevation of the locations, the geological medium underlying the instrument, and the direction of the site from the blast. Observers close to the blast reported sharp vertical motion perception but no noticeable horizontal motion. Farther from the site, the horizontal component became noticeable, but the vertical component was still predominant.

A comprehensive inventory of structures was made for a distance of 35 km from ground zero. Table 4-5 illustrates the distribution of structures and the damage complaints received. Within 15 km (9.35 miles) of ground zero, there was damage to 87 of 282 existing chimneys, to 61 of 95 structures with interior plaster, and to 18 of 35 structures built with masonry walls.

Of the 455 complaints received from 251 locations, 319 were judged by AEC officials as presenting a valid claim for blast-incurred damages. The total amount of payment was $93,831.45 one year later. Residents were allowed one year from the date of the blast to file formal damage complaints; after that period, no claims were accepted for filing.

115

TABLE 4-4
MEASURED PEAK ACCELERATIONS

Site and distance from ground zero	Vertical acceleration (g)	Horizontal acceleration (g)
Eames Orchard—6.99 km	1.70	0.82
Lemon Ranch—6.95 km	0.54	0.20
Grand Valley School—11.43 km	0.60	0.13
Anvil Points—13.83 km	0.20	0.11
Union Carbide Plant—18.12 km	0.10	0.15
Collbran—18.24 km	0.15	0.05
Rifle—20.28 km	0.08	0.08
DeBeque School*—24.63 km	0.09	0.05
DeBeque Business District†	0.21	0.11

*Hard rock.   †Alluvium.
Source: Blume and Associates, Research Division, *Structural Effects of the Rulison Event,* PNE (Dec. 1969).

TABLE 4-5
REPORTED DAMAGE RESULTING FROM THE
RULISON BLAST

Distance (miles)	Number of locations where damage was reported	Total number of locations
0-3.1	1	6
3.1-6.25	54	184
6.25-9.35	87	146
9.35-12.5	17	209
12.5-15.6	84	893
15.6-18.8	3	184
18.8-20.6	5	265
TOTAL	251	1887

Source: L. A. Lee and R. E. Skier, *The Effects of the Rulison Event on Building and other Surface Structures,* Blume and Associates, Research Division (PNE-R-21), Jan. 14, 1970.

The AEC's geological consultants concluded that: "The precautions taken in having people evacuated from the area or outside of the house and two building heights away from the house were well advised as was the care taken during the pre-shot activity in removing or rebuilding chimneys in close-in locations."[55]

The pre-shot estimated cost of repairs for generally superficial damage was approximately $235,000. This estimate did not include the cost of interrupted operations or damage to transportation, utility services, dams, reservoirs, canals or arable lands.[56]

"It is easier to predict damage to structures on qualitative basis (the total amount of damage that should occur to structures) than to predict the actual cost of damage settlement."[57] This situation is illustrated by the difference between actual settlement costs and the predicted damage repair costs. Predicted damage costs are presumably intended to cover all predicted damage, even that which does not occur and that which is not discovered or claimed. It is not likely that the total predicted damage costs will ever be paid in full.

In addition to the survey of structures, a survey of existing wells and mines was also made. The Bureau of Mines concluded that "no surface damage was sustained by any of the gas wells."[58] Preliminary investigations did not find any evidence of subsurface damage either. The wells surveyed were all conventional gas-field wells. As no damage was predicted for these wells, no precautionary measures were taken.

Mine damage results solely from ground motion. The AEC set arbitrary geographic limits on the types of seismic damage expected for administrative purposes. In each geographic area, different precautions were taken because of the expected severity of the seismic shock. The Rulison experience of nonradial propagation of seismic waves extended these arbitrary administrative limits to 37 and 52 miles for peak accelerations of 0.01 g and 0.005 g, respectively.[59] In the area where 0.01 g or greater accelerations were expected, mines were evacuated. In the area where expected accelerations were between 0.01 g and 0.005 g, mine operators were warned to expect the shock but were not required to evacuate. Where accelerations less than 0.005 g were expected, no precautions were taken. Visible mine damage was experienced from peak accelerations of 0.038 g at 27 miles and 0.007 g at 32 miles.[60] However, the blast did not create any apparent damage to the Government Oil Shale Mine at Anvil Points, only 8 miles from ground zero.[61] Because of this discrepancy, Bureau of Mines investigators concluded that "better criteria are obviously needed for relating ground motion to mine damage. . . ."[62] This comment seems particularly appropriate in view of the fact that it is possible that not all of the geologic effects from the Rulison shot were immediately apparent. Damage such as hidden subsurface fractures and strata separations may not show up for long periods of time.

The amount of damage resulting from the seismic shock propagated by the Rulison blast can be considered acceptable for one experiment. It might, however, be unacceptable to the local inhabitants for a prolonged series of shots. Whether it is financially acceptable to the project sponsors is another question. There was no provision made for payment of damages discovered after the one-year time limit set for filing claims. No explanation has been given for the difference between the predicted and actual damage figures. Even though the pre-shot damage figures were conservative, more extensive damage was predicted than was paid for in damage claims. It might be questioned whether adequate compensation has been provided for *all* blast-incurred damages.

Project Rulison did not trigger any earthquakes, and there was no appreciable increase in seismicity in the area. AEC consultants concluded that "the average earthquake (micro earthquake) occurrence of 0.7 earthquakes per day during the two months of recording after the event compared favorably with 0.6 earthquakes per day before the event."[63] The frequency of occurrence as well as the intensity of earthquakes is an indication of the amount of stress built up. Thus, if earthquakes in a particular area increase in frequency and/or intensity, it is an indication that larger and possibly damaging earthquakes may follow. The AEC states in its Environmental Statement for Project Rio Blanco that "it is not impossible in principle for a nuclear explosion, if fired in a region where high natural stress had built up," to trigger "an earthquake with greater seismic energy than the explosion itself."[64] This is possible because the force of the blast-created shock waves may be enough to momentarily reduce the pressure on adjoining layers of rock where stress is built up. If this pressure reduction is large enough to allow slippage of rock layers, due to the increased stress, an earthquake could result. It should be noted, however, that a nuclear device cannot trigger an earthquake which would not have occurred at some time anyway under its own impetus. Colorado is not a very active seismic area, as there are no major faults in the state. There is a strong indication that many of the earthquakes experienced in Colorado were caused by high pressured injection of water into the ground.[65] Such an injection system was used to dispose of wastes at the Rocky Mountain Arsenal near Denver. Since the injection has stopped, earthquakes have been few in number and insignificant in magnitude. It therefore seems unlikely that a nuclear blast would prematurely set off an earthquake in the state.

About one and one-half hours after detonation, water flow in Battlement Creek increased from 4.5 to 14 cubic feet per second. The water in nearby springs, wells, and Battlement Creek increased in turbidity. The water supplies took from 12 hours to 3 days to return to pre-shot water quality. AEC investigators concluded, "Analysis of spring water collected also indicated that the nuclear detonation had no permanent effect on the chemical characteristics of the water."[66] Consultants studying other environmental results concluded that "effects of the Rulison detonation on the surrounding ecosystem were minimal, and no significant adverse effects were noted during the post-event observations."[67] Similarly, rock falls resulting from Rulison-related ground motion were small in size and few in number. There are many natural rock falls in the region because the rocks of the valley walls are intensely fractured, but no extra efforts by highway personnel were required to maintain the road networks in proximity to the blast site.

The small number of rock falls, the fact that there was apparently no damage to the ecosystem or triggering of earthquakes by the Rulison blast does not minimize the seismic dilemma of the AEC. The Rulison shot apparently did not create enough rock fracturing to stimulate gas production economically. Thus, larger total explosive yields, either in a single device or multiple devices in the same well, will be needed in future projects, even though the amount of damage engendered by the Rulison shot is perhaps already too great to be sustained on a repeated basis. Technology may yet find a way out of this dilemma, but that way was not demonstrated by the Rulison experiment.

NOTES

[1]*Denver Post*, Dec. 25, 1969.

[2]Austral Oil Co., and CER Geonuclear Corp., *Summary Project Rulison Feasibility Study*, and *Feasibility Study* (PNE-R-12), Las Vegas, Nevada, 1966. See pp. 2, 6, of the *Summary Feasibility Study* for the figure of over 100 detonations of approximately 200 kilotons each.

[3]*Summary Feasibility Study*, p. 2.

[4]See, for example, *Rifle Telegram*, Oct. 17, 1968. The *Telegram*, later to become one of the Project's harshest critics, at this time, confined itself to simply reporting current developments.

[5]Contract No. AT (26—1)—429.

[6]*Rifle Telegram*, April 10, 1969.

[7]This summary of Project Rulison objectives is taken from both the *Rifle Telegram* article of April 17, 1969, and from *Project Rulison*, Joint Office of Information, May 1, 1969, p. 4.

[8]Joint Office of Information, *Project Rulison*, 1969, p. 3.

[9]The cost figures to November, 1970, were taken from the *Denver Post*, November, 1970, and through May, 1971, from the *Rifle Telegram*, May 31, 1971.

[10]Letter from William H. Carlson of Colorado Interstate Gas Company, to R. E. West of University of Colorado, April 6, 1973.

[11]*Rifle Telegram*, April 17, 1969.

[12]See *Rifle Telegram*, April 24, 1969. For details of the safety program, see: "Project Rulison—Planning Directive," (PNE-R-10), May, 1969, or "Project Rulison Definition Plan," (PNE-R-11), March 26, 1969.

[13]*Ibid.*

[14]*Denver Post*, March 27, 1969.

[15]In a letter to Mr. Ten Eyck on May 5, 1969, Dr. Colburn said, "both the Rulison Project and the Atomic Storage Project have been extensively and scientifically developed. The proponents of both projects have kept the State of Colorado fully informed of their respective projects and should, therefore, not be further delayed by the retroactive application of new rules and regulations." See also: *Official Transcript Proceedings of September 3, 1969*, in the U. S. District Court, for the District of Colorado, p. 66, Civil Action 1722. And in a letter dated May 14, 1969, to Mr. Will Frank of Austral Oil Company, Colburn suggested that ". . . both the Rulison Project and the Atomic Storage Corporation's project be given 'grandfather' considerations because these projects were initiated in the State of Colorado prior to the promulgation of these rules and regulations." (*Transcript*, p. 63.)

[16]*Denver Post*, April 29, 1969.

[17]*Denver Post*, May 7, 1969.

[18]H. Bolton Seed and James L. Sherarc, "Harvey Gap Dam Safety Study," Woodward-Clyde and Associates, May, 1969 (PNE-R-30).

[19]*Ibid.*, p. 10.

[20]*Denver Post*, May 7, 1969.

[21]*Denver Post*, July 17, 1969, editorial. It should be noted that there is no legal means by which the Colorado Department of Public Health can require its approval of nuclear stimulation experimental plans. The Department apparently assumed that Rulison sponsors would abide by their recommendations in respect to public health procedures.

[22]*Denver Post*, July 27, 1969.

[23]*Boulder Daily Camera*, May 6, 1969.

[24]*Boulder Daily Camera*, May 16, 1969.

[25]*Denver Post*, July 28, 1969.

[26]*Ibid.*

[27]The Associated Students of the University of Colorado, Anti-Pollution Committee, "Rulison, Civil Rights and Money," Boulder, Colorado, Aug. 31, 1969.

[28]*Denver Post*, Aug. 13, 1969.

[29]Reported in the *Denver Post*, Aug. 26, 1969 as well as in Court transcripts. The sponsors made the point at this opening session that each day's delay of the scheduled detonation date while the case was in court would cost an additional $16,000.

[30]*Denver Post*, Aug. 21, 1969.

[31]*Denver Post*, Aug. 22, 1969. Rep. Aspinall was chairman of the U.S. House Committee on the Interior. He was defeated in a 1972 primary contest, perhaps in part because his district was redrawn as a result of the 1970 census, his advancing years and the determined opposition of a group of environmentalists who opposed his alleged failure to adequately consider environmental issues.

[32]*Denver Post*, Aug. 31, 1969.

[33]*Denver Post*, Sept. 2, 1969.

[34]*Rifle Telegram*, Sept. 2, 1969.

[35]*Denver Post*, Sept. 11, 1969.

[36]*Denver Post*, Nov. 4, 1974.

[37]*Crowther v. Seaborg*, 312 F. Supp. 1205 (D.C. Colorado, 1970).

[38]The Court acknowledged that Dr. Glenn Seaborg was being sued in his individual capacity, but noted that the suit, in reality, was a suit against the actions of the Atomic Energy Commission as an entity in itself. It was thus assumed that the Project Rulison litigation here was a suit against a governmental agency and dealt with the question of soverign immunity to the extent indicated above.

[39]Survey results reported in the *Grand Junction Daily Sentinel*, May 7, 1970.

[40]See Chapter 7 for discussion of the Rulison Project in the aftermath of the Rio Blanco experiment.

[41]M. Reynolds, "Project Rulison, Summary of Results and Analysis, (PNE-R-55), October 1971.

[42]B. Rubin, L. Schwartz, and D. Montan, *An Analysis of Gas Stimulation Using Nuclear Explosives*, Lawrence Livermore Laboratory Report UCRL-51266, May 15, 1972, p. 19.

[43]Austral Oil Co., Inc., *Project Rulison Summary Feasibility Study*, July 1966.

[44]CER Geonuclear Corp., *Project Rulison Definition Plan*, (PNE-R-11), March 26, 1969.

[45]See, for example, R. E. Duff and L. Schalit, "The Chemistry of the Gasbuggy Chimney," *Nuclear Technology, 11*, (July 1971), p. 390-399.

[46]Roy B. Evans and David E. Bernhardt, *Public Health Evaluation: Project Rulison*, Bureau of Radiological Health, U.S. Dept. of Health, Education and Welfare, (PNE-R-34), May 1970.

[47]P. T. Voegeli, *Geology and Hydrology of the Project Rulison Exploratory Hole, Garfield County, Colorado*, USGS 5-474-16, April 4, 1969.

[48]*Hearings Before the Subcommittee on Air and Water Pollution*, 91st Cong., Aug. 15, 1970, pp. 926, 927. For further information, see Voegeli, *Geology and Hydrology*.

[49]C. F. Smith, Jr., *Gas Analysis Results for Project Rulison Production Testing Samples*, Lawrence Livermore Laboratory Report, (PNE-R-58), Nov. 29, 1971.

[50]W. L. Robison and L. R. Anspaugh, *Assessment of Potential Biological Hazards from Project Rulison*, Lawrence Livermore Laboratory Report, (PNE-R-33), Dec. 18, 1969, pp. 1-30.

[51]L. R. Anspaugh, J. J. Koranda and W. L. Robison, *Environmental Aspects of Natural Gas Stimulation Experiments with Nuclear Devices*, Lawrence Livermore Laboratory Report, (PNE-R-53), Aug. 27, 1971, pp. 1-46.

[52]*Federal Register*, June 9, 1971, pp. 1113-1117.

[53]Peter C. Loux, *Seismic Motion from Project Rulison*, Environmental Research Corp. (PNE-R-18), 1970, pp. 6, 7.

[54]U.S. Bureau of Mines, *Mine and Well Effects Evaluation for Project Rulison*, (PNE-R-39), March 1970.

[55]L. A. Lee and R. E. Skjei, *The Effects of the Rulison Event on Building and other Surface Structures*, Blume and Assoc., Research Division, (PNE-R-21), Jan. 14, 1970, p. 9.

[56]Blume and Assoc., Research Division, Project Rulison: Pre-Shot Investigation and Safety Hazard Evaluations, (JAB-99-61), Aug. 1969, pp. iii, 32, 33.

[57]L. A. Lee and R. E. Skjei, (PNE-R-21), p. 9.

[58]Bureau of Mines, (PNE-R-39), p. 5.

[59]*Ibid.*, p. 20.

[60]*Ibid.*, p. 8.

[61] Hobart A. Merrill *et al.*, *Dynamic and State Responses of the Government Oil Shale Mine at Rifle, Colorado to the Rulison Event*, U. S. Bureau of Mines, Denver, Colorado, (USBM-1001), p. 13.

[62] *Ibid.*

[63] R. Navarro and G. Wyollel, *Rulison Seismic Effects*, Coast and Geodetic Study, Las Vegas, Nevada, (CGS-746-2), Feb. 16, 1970, p. 19.

[64] *Environmental Statement: Rio Blanco Gas Stimulation Project*, (WASH-1519), April 1972, pp. 3-13.

[65] D. M. Evans, "The Denver Area Earthquakes and the Rocky Mountain Arsenal Disposal Well," *The Mountain Geologist, 3*, Nov. 3, 1966, pp. 23-36.

[66] U.S. Department of the Interior Geological Survey, *Geohydrology—Project Rulison*, (PNE-R-24), March 1970.

[67] R. G. Fuller and R. S. Davidson, *Post-Event Bio-environmental Safety Aspects—Project Rulison*, Battelle Memorial Institute (BMI-17-201), Jan. 6, 1970, p. 5.

Chapter
5
Project Rio Blanco:
Its Evolution and Impact

Project Rio Blanco, the last of the AEC's nuclear stimulation projects, was carried out on May 17, 1973. The story of Rio Blanco is not yet complete because evaluation and accumulation of data and analysis of test results are still in progress. Although a final assessment of the project has yet to be made, the history of Rio Blanco already includes much that is new to the Plowshare Program in political, legal, and economic ramifications. Perhaps the most important impact of the project has been the heightened public consciousness of nuclear stimulation in Colorado.

As in previous Plowshare projects, the AEC combined with private firms to design and execute this most recent nuclear stimulation experiment. Cosponsors for the Rio Blanco Project include the Equity Oil Company of Salt Lake City, Utah, and the by now familiar CER Geonuclear Corporation of Las Vegas, Nevada. The basic purpose of the test, as in the earlier Gasbuggy and Rulison efforts, was to free natural gas from tightly compacted sandstone formations, which in this case underlie the Piceance Creek Basin in northwestern Colorado. The sponsors considered the Rio Blanco experiment a next step, vital to perfecting the technology, after the experience gained from the single nuclear shots used in Gasbuggy and Rulison. Three newly developed 30-kiloton Miniata nuclear devices were used in an effort to create three interconnecting chimneys into which gas could flow and then be brought to the surface. Should this three-device test prove successful in Phase I of the experiment, the sponsors envisioned detonating as many as 350 nuclear devices in follow-up phases of the project, leading ultimately to full field development.

These plans became matters of governmental and public concern at a time when the energy crisis began to make itself felt in Colorado. During the winter of 1973, just prior to the Rio Blanco detonation, Denver public schools were briefly forced to curtail the school week because of inability to heat school buildings. The following winter, many rural homes were inadequately heated because of the shortage and rapidly rising price of propane gas. The energy crisis was no longer a somewhat vague possibility for the future. In Colorado, as elsewhere, it was here and now, and all too real.

Yet in spite of the obvious need for increased energy supply and distribution, there were other forces at work which precluded smooth sailing for Project Rio Blanco. The limited, though highly vocal, outcry over the Rulison Project, the increasing public questioning of the government's handling of nuclear stimulation, and the ensuing litigation combined to form the backdrop against which the Rio Blanco story would be played out. The question of whether nuclear stimulation experiments should be continued in the state became again a highly political issue, reflected generally in the campaigns of aspiring political leaders, and specifically in a referred proposal placed on the 1974 general election ballot.[1] This proposal stated that further nuclear explosions in the state would be prohibited without a preceding specific vote of approval by the people. The citizens of Colorado adopted the proposal by a wide margin, and in doing so made it abundantly clear that they wished to be actively consulted on future nuclear operations which could affect them and their environment. Whether the voters will have the final voice in the decision-making process on the use of nuclear explosives in the state remains to be seen. It is possible that the courts may once again be called in as the final arbiter of potentially conflicting governmental, industrial, and citizen rights and prerogatives.

How and why did the citizens of Colorado become so involved in the conduct of nuclear stimulation technology? The answer is not

126

totally clear, but it is obvious that as Project Rio Blanco evolved, not only local, state, and federal government officials but also political leaders, the courts, and citizens became increasingly concerned about the conflict between increasing energy supplies and providing adequate environmental protection.

We now turn to a study of this evolution, first by tracing political, governmental, and social reactions to the project, then by considering various technical and scientific factors, and finally by examining some legal and related elements of the picture as they appeared to be in the fall of 1974. Though retracing the steps and processes leading to the Rio Blanco detonation will not provide a final answer to the future of nuclear stimulation projects, it will throw some light on how and why decisions to proceed were made, and also possibly why the quality of life that the people of the state of Colorado wish to create and perpetuate for themselves and future generations became a basic issue.

PRELIMINARY COLORADO REACTIONS: OFFICIAL AND UNOFFICIAL

In contrast to the situation in the years preceding the Rulison detonation, Governor Love made an effort to play a much more active role early in the Rio Blanco decision-making process. Probably as a result of growing environmental awareness and concern about the effects of nuclear experiments on potential oil-shale development in Colorado, the Governor tried to secure a broad spectrum of opinion on the Rio Blanco proposal. In a related, but equally significant effort, he tried to obtain assurances from the AEC that he could veto the use of nuclear explosives in Colorado if he felt that their use would endanger the health and welfare of the people in the state. In his first effort the Governor was successful. He received more information, advice, and varied opinions prior to Rio Blanco than prior to Rulison— possibly because there was at that time more information on nuclear-stimulation technology available. In his second effort, the Governor was less successful. It became apparent that he could not, or did not, have the power to veto underground nuclear blasting in the state, even if he believed that this would be the most desirable course to follow.

Perhaps to avoid a repetition of the kind of public concern and criticism that arose over the Rulison Project, the Governor appointed an 18-member advisory group to look into the plan for the Rio Blanco Project and to advise him on its acceptability. The group was chosen to reflect both local and statewide viewpoints, as well as the perspectives of those concerned with technology, industry, government, and the environment. In view of later vigorous protest by medical doctors opposed to the project, it is noteworthy that there was no practicing radiologist or member of the medical profession, other than the Chief of Occupational Radiological Health Division of the Colorado State Department of Public Health, on the Advisory Group. (Due to expressed concern by members of the medical profession in Colorado, a practicing radiologist was later added to the Advisory Commission.) The Advisory Group was appointed in June 1971, and at that time it was announced that a majority and minority report would be prepared for the Governor's information.

Throughout its investigations, the Governor's Advisory Committee held meetings with the sponsors of Project Rio Blanco as well as with those who were opposed to the nuclear detonation. In June 1971, officials from the CER Geonuclear Corporation met with the Committee to familiarize the members with basic Project Rio Blanco plans and concepts. A month later, there was another briefing in Grand Junction, Colorado, which was followed by a tour of the project test area in the Piceance Creek Basin.

In meetings that were held in October 1971, The Oil Shale Company (TOSCO), Shell Oil Company, and Wolf Ridge Minerals expressed strong opposition to the use of nuclear stimulation techniques in the Rio Blanco field. In general, companies interested in oil-shale development felt that nuclear stimulation of natural gas and oil-shale mining are not compatible technologies when employed simultaneously for the recovery of geologic resources located in the same area. Both oil-shale developers and companies holding other mineral leases in the Rio Blanco field maintained that nuclear blasts would disrupt overlying geological strata and consequently make it impossible, or at least more difficult, to mine either the oil shale or other mineral resources which lie above the gas-bearing formations. TOSCO representatives repeatedly stated that their company had no objection to nuclear stimulation in principle, but objected to the use of this technology in Project Rio Blanco. They argued before the Governor's Advisory Committee and elsewhere that the Rio Blanco field contains far more potentially valuable oil in shale than extractable natural gas. If Phase I were permitted and then followed by full field exploitation, oil-shale development would be unreasonably postponed, an action which would be to the ultimate detriment of both the economy of the state of Colorado and the total available energy supplies of the United States.

Members of CER, Equity Oil Company, and the AEC had the opportunity of contesting the oil-shale industry objections in December 1971, when further meetings were held with the Advisory Committee in Las Vegas. Committee members, members of the press and representatives of the oil-shale industry were invited to the meetings and a tour of the AEC Nevada Test Site, where much preliminary testing of Plowshare technology had been carried out.

In order to gain some insight into local sentiment regarding Project Rio Blanco, members of the Advisory Committee attended, in January of 1972, a public meeting in Meeker, the town nearest the proposed test site. This meeting, sponsored by the Rio Blanco Planning Commission, was called to provide information on the project and to obtain a sampling of local reactions and attitudes about it. Individual members of the Committee also attended another meeting in February 1972, in which officials of the AEC and CER discussed the project with an audience of approximately 125 people. This group, composed largely of environmentalists, expressed serious reservations about the wisdom of proceeding with the Rio Blanco plans. There was considerable discussion as to whether nuclear-stimulation technology would significantly increase the national fuel supply and whether the amount of gas produced would be sufficient to warrant the potential environmental risk involved.

The Advisory Committee, often considering varying opinion and testimony and doing some independent research, presented its report to Governor Love in late February of 1972. The Governor then simply announced publicly that a majority of Committee members favored proceeding with Phase I, but that this preliminary phase of the project should be undertaken only subject to certain specific conditions. When these stipulations were released in July 1972, Love noted that the Committee's report was not a binding one, but only an advisory recommendation. Before making any final executive decisions on the project, he said he would consider studies made by the Interior Department on the compatability of nuclear-stimulation technology and oil-shale development in the same geographic area and geological fields.[2] Specific conditions recommended by the Advisory Committee included the following prerequisites:

> 1. The state would obtain a formal agreement from the U.S. Atomic Energy Commission under which the Governor of Colorado would be kept "fully and currently informed" on the progress of all plans and operations in Project Rio Blanco, and under which he would have the power to veto any proposed detonations of nuclear explosives.
>
> 2. The U.S. Atomic Energy Commission would be required to enter into a formal agreement with the State of Colorado, under which it would acknowledge complete ownership and liability in perpetuity for all post-detonation fission products remaining underground, and a further formal, legal agreement clause that the Atomic Energy Commission would be the agency responsible for the removal or disposition of any misfired nuclear device, subject to the state's approval of such disposal plans if they should ever be necessarily put into effect.
>
> 3. The State of Colorado would require accompanying experiments or at least joint monitoring and tests, designed to demonstrate clearly the effects of nuclear stimulation of natural gas resources on any oil-shale deposits and on any water-bearing rock formations, to find if there are any possible or actual adverse effects on future oil-shale development or the ground-water system from nuclear-stimulation projects.
>
> 4. The state would require that no water-contained radioactivity be introduced into an area where biological life exists.
>
> 5. The State of Colorado would make "perfectly clear" that there is no state approval implied of any additional plans or phases of Project Rio Blanco from a recommendation by the Governor's Advisory Committee on Project Rio Blanco that Phase I be permitted to continue.

For some still unexplained reason, the full text of the Advisory Committee's report was not made public until five months after it was first presented to the Governor, although the tenor of the recommended stipulations seemed to be in tune with actions that the Governor had previously undertaken.

The history of the Governor's attempts to secure a state-level veto power over nuclear stimulation detonations dates back to a time shortly after the Rulison shot. In a letter to then Chairman Seaborg of the AEC in early 1970, he requested clarification of the state's authority to either allow or disallow such projects. Specifically, he requested assurance from the AEC that it would seek concurrence from appropriate state officials prior to any nuclear detonation in the Rio Blanco field.[3] Dr. Seaborg replied to the Governor's inquiry on December 17, 1970, indicating that the AEC would indeed seek such concurrence prior to final implementation of the project. On the basis of this reply, the Governor apparently assumed through the formative stages of project planning that he had a veto power.

Governor Love was not alone among state political figures concerned about a decision to carry out the project. United States Senator Gordon Allott noted his concern by insisting in a speech before the Senate Subcommittee on Public Works in May 1971 that no contracts for nuclear stimulation projects in Colorado be signed until public meetings were held. Dr. Seaborg promised to hold such public hearings. A year later, in the spring of 1972, Senator Allott asked Rogers C. B. Morton, Secretary of the Interior Department, to defer any decision on detonation until a "detailed report" was made on the possible effects of nuclear-stimulation technology on oil-shale and water resources.[4] It is not clear whether Secretary Morton or the Department of the Interior had binding and final decision-making power, but as a coordinate federal agency, the Interior Department certainly has the authority and responsibility to make studies and recommendations on any matter related to energy resources and development. Senator Allott, with an election battle looming in November 1972, was apparently proceeding carefully in an attempt not to antagonize either pro- or anti-Rio Blanco forces. Although his attempts to postpone a definitive decision on the project were successful, his bid for re-election failed, and he lost his Senate seat to Floyd K. Haskell at the polls in 1972.

131

THE AEC AS ACTIVE PARTICIPANT

Although the AEC supplied the nuclear devices and supervised safety services in all three Plowshare experiments, its role in the Rio Blanco Project took on some added dimensions. Political and legal factors arose in the Rio Blanco Project that did not occur in the earlier Gasbuggy and Rulison Projects. As a result of the passage of the National Environmental Policy Act of 1969, which became effective in January 1970, and the subsequent *Calvert Cliffs* court decision, the AEC was required to thoroughly assess in advance the environmental impact of any project it proposed to undertake.[5]

Under the terms of the act, which is usually referred to as NEPA, it is the policy of the "federal government, in cooperation with state and local governments and other concerned public and private organizations, to use all practicable means and measures, including financial and technical assistance, in a manner calculated to foster and promote the general conditions under which man and nature can exist in productive harmony and fulfill the social, economic and other requirements of present and future generations of Americans."[6] A paramount provision of NEPA is the requirement that a detailed environmental impact statement be made for every major federal project which will significantly affect the quality of the human environment. Such a statement is to be prepared under the authorship of an official of the lead agency which is primarily responsible for the proposed actions. In preparing an environmental impact statement, the lead agency is expected to consult with any other federal agency which has either jurisdiction or special expertise in the particular subject area.

The AEC acknowledged the existence of the new Act by issuing an amendment to its licensing regulations. In the preamble to the text of the new regulations, the AEC made clear that its responsibilities under NEPA would be carried out in a manner consistent with a policy of expediting decisions and avoiding delays which would be detrimental to the national interest.[7] From the AEC point of view, one of the first priorities was, and would remain, the harnessing of the atom in pursuit of increased sources of energy.

The revised AEC regulations did not, however, meet the requirements of NEPA in the opinion of some who opposed the construction of the Calvert Cliffs nuclear power plant in Maryland and a group, known as the Calvert Cliffs Coordinating Committee, challenged the AEC in Federal District Court.[8] At issue was the regulation that the Atomic Safety and Licensing Board, which has the authority to grant or withhold a license for construction of a nuclear power station, may not consider the environmental impact of an application unless a specific environmental question is raised at the Board hearing by a party to the proceedings.

The Calvert Cliffs Committee also contested in court the following AEC regulations which in their view negated the intent of NEPA: (a) no nonradiological environmental issue could be raised if notice for an application hearing was published in the *Federal Register* prior to March 4, 1971; (b) the Licensing Board could not independently investigate determinations made by other agencies; and finally (c) if a construction permit was issued prior to NEPA's effective date, environmental factors would not be considered until the hearing on granting the operating license is held.

Judge Skelly Wright, who heard the *Calvert Cliffs* case in Maryland District Court, was apparently far from pleased with the AEC's performance when he noted that "the Commission's crabbed interpretation of NEPA makes a mockery of the Act."[9] Judge Wright also questioned the AEC's reluctance to meet NEPA's procedural obligations, and said that "whether or not the spectre of a national power crisis is as real as the Commission apparently believes, it must not be used to create a blackout of environmental consideration in the agency review process."[10] The Judge, in short, ruled for the plaintiffs and in doing so made clear that the NEPA mandate requires conscientious and informed consideration of the environmental implications of any proposed AEC project.

As a result of this decision, the AEC had no alternative but to prepare new regulations which would more closely correspond to the need for environmental protection. In September 1971, a new Appendix D to AEC regulations was published in the *Federal Register*. The new regulations were intended to be "responsive to the conservation and environmental concerns of the public [while at the same time] reconciling a proper regard for the environment with the necessity for meeting the nation's growing requirements on a timely basis."[11]

The section of Appendix D of the new regulations which has specific application to the Rio Blanco Project includes the provisions that an applicant for an AEC license must draft an environmental impact statement which must be submitted along with the application. This statement must be publicly available, and the AEC must review it before preparing its own environmental impact report for public circulation and comment. After receiving comments from federal, state, and local agencies, the AEC is required to prepare an environmental impact statement which incorporates the comments, criticism, and suggestions made in the review process. This final statement must also include consideration of alternatives to the methods proposed, an analysis of any irretrievable commitment of resources, and a preliminary cost-benefit breakdown for any proposed project.[12]

Such were the requirements which the AEC faced at the time the Rio Blanco Project was under serious consideration. In some respects, the AEC went beyond the scope of NEPA and its own guidelines, because in early March 1972, an announcement was made that formal public hearings would be held on the project on March 24 and 27 in Meeker and Denver, respectively. Presumably, the purpose of these meetings, which were unprecedented in AEC history, was to give the AEC an opportunity to consider public reaction to the impact statement and to the project itself. Although the purpose may have been to obtain widespread public reaction to the project, it is questionable if this objective was effectively implemented because of the rigorous requirements which an ordinary citizen had to fulfill before being allowed to testify.

Anyone wishing to participate in these hearings was required to file a request with the AEC by March 13, 1972, although this deadline was later extended to accommodate those who had not met it. Each application to participate in the hearings was to include a "resume of the individual's or organization's technical or other qualifications to make a contribution to the proceedings, a position regarding the draft environmental statement and the Rio Blanco experiment and a description of the points he seeks to

134

present. . . ."[13] The applicant was also to state whether he would be represented by counsel. A person who wished to make a limited appearance, simply to present some specific information, would be allowed to do so, provided he met the same requirements as those outlined for "a participant." Persons who could show reasonable cause for not registering in advance also would be allowed to testify at the hearings either as a participant or in a limited appearance. There was no advance indication as to whether an individual could present the same testimony in both Meeker and Denver.

A three-member board convened the hearings in Meeker at 10:00 a.m. on March 24. The three members, chosen by the AEC, included William Mitchell, an attorney from Washington, D.C., Dr. Lee Aamodt, a physicist from the AEC's Los Alamos Scientific Laboratories, and Dr. Max Zelle, a faculty member and chairman of the Department of Radiology and Radiation Biology at Colorado State University in Fort Collins, Colorado. The board was to accept testimony, question participants and those making limited appearances, and ultimately to prepare a summary of the issues raised and public reaction to Project Rio Blanco for consideration by the AEC.

The first testimony presented in Meeker was from representatives of the AEC and the two private contracting companies, CER Geonuclear and Equity Oil Corporations. Their presentation centered on the basic nature of the project, its purpose, and the plans for its implementation. Only the sponsors of Project Rio Blanco presented duplicate testimony of this kind at both the Meeker and Denver hearings. Other parties' testimony could be given at one place or the other, but not both. The representatives from TOSCO, The Oil Shale Company, took immediate exception to this procedure, claiming that it was basically unfair, and made a mockery of a process that was supposed to make information on both sides of the question available to the public in the locale of each hearing.[14] The sponsors considered the objections raised by TOSCO, and at this point proposed that the Hearing Board accept testimony by *all* participants at one place or another, but not both. TOSCO representatives expressed disagreement with this compromise proposal on the grounds that in the absence of testimony in both locations, press coverage would be curtailed, and the result would be less opportunity for public understanding of the issues.

The Hearing Board ultimately overruled TOSCO and accepted the sponsors' solution to the problem, basically on the grounds that the hearings should not be unduly prolonged by repetition and that, in any case, all testimony would be published eventually as part of the public record.[15] Clearly, there were two differing attitudes about the purpose of the hearings. TOSCO wanted as wide a dissemination of viewpoints as possible, and the AEC wanted expeditious hearings. The hearings, though not required by law, did fulfill a promise made by Chairman Seaborg to Senator Allott that public hearings would be held, and as such were probably not part of original Rio Blanco planning. The target date for completion of Phase I varied, but time was marching on, and the sponsors wished to move ahead with the least possible delay. TOSCO, on the other hand, wished to delay or indefinitely postpone the Project, thereby giving oil-shale recovery operations first priority in the Rio Blanco field.

As a result of the Board's decision not to duplicate testimony, the hearing at Meeker and the one at Denver were different in both substance and mood. At Meeker, four of the six participants were project representatives, and their testimony dealt with the project from engineering, technical, geological, radiological, and safety perspectives. This information was generally presented within the framework of the vital need for increasing gas supplies and the significant contribution that Project Rio Blanco would make toward this end. Of the other two participants, only Frank Cooley, a lawyer and geologist, presented an adverse view. His objections generally centered on grounds of questionable human and environmental safety and potentially dangerous seismic effects. Ronald Gitchell, who spoke for the Rio Blanco County Planning Commission, did not take a stand for or against the project, but urged the AEC to make a thorough evaluation of all parts of Phase I before proceeding with the detonation.

Thirteen people presented testimony in a "limited appearance" category, but only four were residents of the local Meeker area. It is possible that in Meeker, local citizens felt their views were adequately expressed by representatives of the Chamber of Commerce and the Oil Shale Regional Planning Commission, in addition to Mr. Cooley and Mr. Gitchell; at the Denver hearings, three days later, the observation was made that the lack of local participation arose from the AEC's methods of procedure which effectively discourage such contributions.

136

In contrast to the Meeker sessions, the Denver hearing predominantly involved those who opposed the Project for commercial or environmental reasons. Chief among these opponents were representatives from TOSCO, the Superior Oil Company, the Colorado Committee for Environmental Information, Environmental Action of Colorado, and the Colorado Open Space Council. At the time of the hearings, it seemed that the most influential and perhaps damaging testimony on the project would come from oil-shale interests which were presented in Denver by Louis Yardumian, the Vice-President of TOSCO. Neither of the preceding nuclear stimulation projects had been undertaken in the context of concerted opposition by another influential sector of private industry which was concerned with efforts to exploit energy resources in an economically depressed area. This kind of opposition could not in any way be characterized as "obstructionism by fuzzy-headed environmentalists."

Mr. Yardumian presented the oil-shale industry's case. He estimated the value of oil-shale deposits in the Rio Blanco field to be one hundred times larger than the natural gas resources in the same area. In his view, it would be extremely unwise to proceed with a nuclear explosion which could effectively destroy the opportunity to exploit overlying oil-shale formations. The sponsors, for their part, had repeatedly stated their conviction that the nuclear detonation would occur at such great geologic depth that the overlying oil-bearing formations would not be disturbed. Mr. Yardumian, however, was critical of the fact that the AEC was the lead agency which had the responsibility of preparing and filing the Rio Blanco environmental impact statement. He maintained that in a matter which so vitally affects natural resources, the U.S. Department of the Interior should have had prime responsibility for the project. The question of ultimate federal agency responsibility was a new ingredient in AEC Plowshare experience, because its primary and complete responsibility for all things nuclear had never before been publicly questioned in this manner.

Dr. H. Peter Metzger, an outspoken environmentalist, physicist, science journalist, and a member of Governor Love's Advisory Committee on Project Rio Blanco, presented the most emotional appeal against any nuclear stimulation shot in the Rio Blanco field. He criticized the AEC for the overly formal and rigorous manner in which the hearings were conducted, contending that the requirements frightened off Rio Blanco area residents who opposed the blast, but were too intimidated to say so. Dr. Metzger further argued that it was both illogical and wholly misleading to consider Project Rio Blanco simply as a one-shot experiment, since up to 560 detonations would be required to fully exploit the gas in the Rio Blanco field, and up to 13,000 nuclear detonations would be needed to develop all the natural gas in tightly compacted geological formations in the Rocky Mountain West. He basically saw no sense in embarking on an experiment which even if successful in producing usable gas (which was unlikely in his opinion) could not conceivably be used on a large enough scale for a viable commercial operation. He regarded the prolonged and frequent nuclear blasting that would be necessary for optimum use of the technology as unthinkable. Intrinsic to this argument was Dr. Metzger's view that uranium used in the Rio Blanco devices will be in increasingly short supply and that eventually plutonium, a much more polluting substance, will have to be substituted for it. The result could only be increased biological and environmental hazard. As Dr. Metzger saw it, Colorado residents could not be expected to endure a great number of nuclear blasts for the dubious pleasure of receiving increasingly dangerous radiation exposure.

Dr. Edward Martell, an atmospheric geochemist at the National Center for Atmospheric Research in Boulder, and a representative of the Colorado Committee for Environmental Information, found fault with various sections of the draft environmental impact statement. Among other things, Dr. Martell questioned the conclusion that there would be no significant radiation danger to either subsurface water or local residents. For essentially the same reasons that CCEI had opposed Project Rulison, it now opposed Project Rio Blanco. In addition, Dr. Martell faulted the draft impact statement for its failure to deal adequately with alternatives to nuclear stimulation technology. He specifically cited the AEC's failure to consider, or even mention, new developments and techniques in hydraulic fracturing methods. He also criticized the AEC's preliminary cost-benefit analysis which did not include the cost of the energy required to produce the nuclear materials, the losses of fissionable materials to other nuclear power sources such as nuclear reactors, and the loss of natural gas through the blasting and subsequent flaring operations.

David M. Evans, a geologist and director of the Potential Gas Agency at the Colorado School of Mines in Golden, spoke as a representative of the Colorado Open Space Council and the Audubon Society. He too questioned the AEC's estimated number of explosions which would be necessary for full field development of gas in tightly compacted geological formations in the Rocky Mountain West. Also, his calculations indicated that approximately 13,000 nuclear explosions would be necessary to free such trapped gas in western states. Mr. Evans' prime concern, however, was that multiple detonations of nuclear bombs beneath the Upper Colorado River Basin might release radionuclides into the water supply of the entire area. Should such contamination occur, he predicted that long-lived isotopes would produce irreparable environmental damage for a period exceeding *six hundred years*, to say nothing of the consequences to human life in a far shorter time span. In short, he questioned the sponsors' claim that all blast-created radioactivity would be contained in a resolidified "melt" at the bottom of the nuclear chimney.

Mr. Evans also requested information from the AEC and CER on who would be financially responsible for damage to property as a result of the blasting. Mr. Herbert Grier, President of CER, replied that insurance contracts were being negotiated but that at the moment he could give no definite figures on how much of the damage costs would be absorbed by private industry. The contract as finally written provided that CER was to obtain $10 million insurance for damage claims, which were estimated to be no more than a possible $64,000. If damage claims exceeded that amount, the federal government was to assume the additional costs.

Although at the Denver hearing much negative testimony was presented, there were a few speakers who approved the project. Among the proponents was Mr. John Rold, the Chief Geologist for the State of Colorado, who had previously endorsed the operating procedures and geological safety of the Rulison experiment. Mr. Rold had participated in the planning of the analysis and documentation of expected ground-motion effects in the Rulison Project and was similarly involved in Project Rio

139

Blanco in his official capacity as State Geologist. On the basis of his studies, he again found from a geological point of view no unacceptable risks in the use of nuclear-stimulation technology in Colorado.

The Meeker and Denver hearings clearly provided opportunities for airing of divergent views on Project Rio Blanco. Yet, after the conclusion of both meetings, and with no further convening of the special Hearing Board, it was still not certain whether the primary purpose of the March days of testimony was to receive data in succinct form, or to provide an opportunity for widespread public participation and evaluation of both sides of the question. If the original intent of the hearings was to open up the early stages of Plowshare Program projects planning to public participation, this purpose was not well fulfilled. The rigorous requirements for participation and the last-minute procedural ground rules precluding similar testimony at both Meeker and Denver hindered public participation and exposure.

On the other hand, in the very fact of setting up procedures for public hearings, the AEC did carry out a new function. Clearly, there must be some means of limiting debate, even if the broadest view of participatory democracy is taken. However, the powers of the Hearing Board were not precisely defined; if the function of the Board was to moderate sessions and present testimony for public information, this was done. On the other hand, if the Board was to function in the manner of a citizen review board, the selection of its members, its authorized authority, and its method of procedure left much to be desired. The Board members themselves, perhaps, had differing views of their functions. Unlike a labor arbitration board or panel of courtroom judges, no summary of individual board member opinion was made public. Ideally, some middle ground should be found between the government's right and responsibility to make informed and expeditious decisions and the citizens' need to participate in decisions which inevitably and vitally affect them. In retrospect, the Meeker and Denver hearings could be considered halting steps in this direction.

THE WESTERN SLOPE SCENE

Although the local population in the vicinity of the Project Rio Blanco test site did not turn out in force to testify at the AEC's Meeker hearing, the people of the area did receive a great deal of information from the sponsors about the proposed project at public meetings and through the local newspapers. The first of a series of local meetings, held in January 1971, at the Piceance Creek School, at Meeker, and at Rangely, were designed to inform the people that a nuclear-stimulation project in the Piceance Creek Basin was under active consideration. A year later, a second series of meetings took place at Meeker, Piceance Creek, and DeBeque. The Meeker meeting was attended by 250 people, including several members of the Governor's Advisory Commission. The purpose of these meetings on the Western Slope was to present a detailed explanation of nuclear-stimulation technology and the prospects for full field development of the Rio Blanco unit by representatives of the AEC, CER, and Equity Oil Company. Very few questions were asked, but on the basis of various comments, the *Meeker Herald* reported that the audience seemed most concerned about the prospect of a large number of blasts, damage to the saline water table and surface springs, and the danger from radioactivity due to flaring of the gas.[16]

Local residents of Meeker quite naturally were much interested in the physical and economic consequences of Project Rio Blanco for the area. In February 1972, Hal Aronson, a vice-president of CER, gave projected employment figures for the project at a Rio Blanco County Planning Commission meeting. Initially, he said, 15 employees would be needed at various intervals during demonstration phases of the project, and a maximum of 250 workers would be hired during full field development. At most times, however, a staff of 75 to 100 would be adequate for the project. After completion of full field development, a regular crew of only about 12 people would be necessary for routine maintenance purposes. Mr. Aronson further estimated that the average assessed valuation of the Rio Blanco gas field would be $32.4 million. According to his calculations, this total valuation would generate about $1 million in state taxes each year, and about $1.5 million in Rio Blanco County taxes at the average 1970 county mill levy of 48.42 mills. Total tax collections in the county from property taxes amounted to approximately $2.8 million in 1970.[17] According to Mr. Aronson's projections, local residents could anticipate a substantial broadening of the tax base if full Rio Blanco field development were undertaken.

In contrast to Projects Gasbuggy and Rulison in which no specific legal permission from local governmental units was required, it was necessary to obtain a special use permit from the Rio Blanco County Planning Commission to conduct the Rio Blanco experiment. Most of the county is zoned for agricultural use. By county regulation, any experimental mineral use of the land other than the drilling of wildcat wells requires explicit permission by the Planning Commission. The sponsors of Project Rio Blanco initially claimed that their plans were essentially for a "sophisticated wildcat well," and that no permit was required. However, at least one member of the Commission stoutly maintained that no matter what the decision of the AEC, other sponsors, the Governor or state agencies, the Rio Blanco Project could not proceed without formal affirmative action by the Commission. CER eventually agreed to meet the Commission's requirements and applied for a special use permit.

Rio Blanco county regulations specifically state that no permit will be granted to a project which will cause pollution of air or water. In applying for the special use permit, CER referred to the AEC's environmental impact statement, acknowledging that some pollution of the air might occur from gas flaring during production. This did not appear to be a serious problem to the majority of the Commission, and the application was approved for a two-year period on April 27, 1972, by a 4 to 1 vote. After April 1974, presumably there would be routine production of natural gas from the project which would not require a special use permit from local authorities. In all probability, the claimed restraining power of the locality as exercised by the Planning Commission could never be effectively enforced, but the question became academic following the issuance of the special permit.

The picture of the Planning Commission standing up for its assumed rights against nation, state, and industry is indeed interesting in the complex of interrelationships and interests involved in nuclear experimentation. If the people directly affected by the consequences of a developing technology are to exert any substantive influence on the decision-making process,

142

that often nebulous and illusive quantity called public opinion must be identified as accurately as possible. Although extensive efforts were made by the Rio Blanco sponsors to inform the local citizenry about the project, there was no equivalent effort made to determine how much in fact the people really understood. Did the people of Meeker, who were literally and figuratively at the center of the Rio Blanco controversy, approve or disapprove of the project? How intense were their convictions and what were the reasons for the opinions they had?

These questions were investigated by an interdisciplinary research group at the University of Colorado which studied the AEC's Plowshare Program and the interrelated web of technical, political, and legal ramifications which it involves. To answer these questions, a systematic survey of opinion about Project Rio Blanco was begun in various Western Slope communities in the fall of 1971 and completed in February of the following year.[18]

A preliminary questionnaire was developed in mid-November of 1971 and pretested in Meeker to determine possible ambiguities in the wording of questions. Revisions were made on the basis of the pretest, and a random sample of Meeker area residents was drawn from telephone book listings.[19] At the end of November, 421 questionnaires were sent out; 181 questionnaires were completed and returned, providing a response rate of 43 percent.

The questionnaire was designed to test the intensity of opinion about the initial blast as originally scheduled for 1972, as well as to poll opinions about long-term full field development. The results were as follows: 21.5 percent strongly favored the 1972 blast, 28.5 percent favored it moderately, 13.4 percent moderately opposed it, and 36.6 percent opposed it strongly. Disregarding the factor of intensity of opinion, 50 percent of the respondents favored the detonation and 50 percent opposed it. It should be noted, however, that the greatest number of respondents were in the "strongly opposed" category. There was, as it turned out, an extremely strong correlation in attitudes toward the initial blast and toward full field development.

The respondents gave a number of reasons for their attitudes. 75.9 percent felt that the Meeker area would benefit economically, but at the same time, 55.4 percent thought that the blasting might contaminate underground water supplies. 52.4 percent believed that not enough is known about the nuclear technology and that on this basis the nuclear shot should not be permitted. Respondents in each of these last two groups were strongly inclined to oppose the project. 51.4 percent of the respondents expected that the blasting would damage buildings in Meeker and the surrounding area, and 49.2 percent expressed fear that it would interfere with the development of the oil-shale industry in the Rio Blanco field.

Attitudes in general could be grouped into two main categories. One group was distinctly apprehensive about possible hazards associated with the blasting, and was therefore likely to oppose the project. A second group anticipated economic benefit from the blasting and thus tended to support the project.

The responses to questions designed to test the effectiveness of the dissemination of information about the project showed a high degree of public awareness and information. Of those responding, 99.4 percent indicated they were familiar with the proposal for underground nuclear blasting at Rio Blanco. 85.5 percent knew that the purpose was to stimulate natural gas production. Although 13.1 percent thought that the purpose was to develop oil shale, 60.7 percent correctly identified CER Geonuclear and the Equity Oil Company as the commercial sponsors. 38.7 percent of the respondents knew that 100 or more blasts would be required for full Rio Blanco field development; 34.1 percent were not aware of the total number, and 13.3 percent believed that only three detonations would be necessary. The remaining 13.9 percent of respondents were divided in their beliefs that full field development would require 1, 10, or 50 blasts.

In correlating attitudes toward the project with the data on those who had accurate information about the required number of blasts, it was found that those with accurate knowledge tended to oppose the project and to worry about associated possible hazards. No relationship was found between accuracy of knowledge about the number of blasts and the anticipation of economic gain. Nor were those who were able to correctly identify the purpose or sponsors more likely to support or oppose the project or to fear hazards or expect economic gain. It was also found that the respondents' opinions toward the project were essentially unrelated to their level of education, present occupation, age, average family income, marital status, number of children, or sex.

Even though this questionnaire was circulated prior to the major series of public meetings subsequently held by the sponsors in the Rio Blanco area, the results show that in December 1971, there was already widespread knowledge about the proposed nuclear detonations and that there was substantial understanding of the details by the people in the area. The survey results also made clear the fact that intensely held opinions, both for and against the project, had already been formed.

To ascertain if the attitudes of Meeker residents were unique or were shared by people elsewhere on the Western Slope, surveys were undertaken in February and March of 1972 in both Rifle and Delta, two other Western Slope communities of approximately the same size and socioeconomic characteristics as Meeker.[20] The town of Rifle, Colorado, was selected because it had previously experienced a similar shot in the Rulison Project, and it is, like Meeker, economically depressed. The town of Delta, Colorado, was selected because it was far enough from either blast site as to be relatively unaffected by the nuclear-stimulation projects.

As in the Meeker survey, the samples were drawn from telephone directories.[21] On February 29, 1972, 431 questionnaires were sent to Rifle; 186, or 43.2 percent, were returned. From the 409 questionnaires sent to Delta residents, a somewhat lower return rate of 129, or 31.5 percent of the total, was obtained. Because the response rate was relatively low, the results do not necessarily represent a random sample of the population. It is interesting, however, that Delta, the town whose inhabitants had never experienced an underground nuclear shot and do not anticipate doing so, showed the lowest response rate. The issue of nuclear stimulation is apparently far more salient to people in locations close to the ground zero point than the people well removed from it.

In comparing returns from the three cities, it was found that there were no statistically significant differences between the three major populations in average educational and occupational status, yearly income, marital status, sex composition, and average number of children. There were, however, some differences in attitude toward the blast, apprehension about it, and expectation of material gain. Further examination of these results revealed that responses from Rifle and Meeker were quite similar, while those from Delta diverged significantly from the

145

other two. As a result of this finding, Meeker and Rifle questionnaires were combined as though coming from one population and were contrasted with those returned from Delta. In this way the entire sample was dichotomized into those cities which had direct exposure to some phase of nuclear stimulation and a city which had none.

Tests were run to determine if there were statistical differences between the two groupings in respect to knowledge, attitude, apprehension, expectation of economic gain, and marital status. Of greatest interest was the finding that the Meeker-Rifle respondents were significantly more opposed to the proposed Rio Blanco blast than were the respondents from Delta. The specific response rate is shown in the following table:

	Strongly in favor	Moderately in favor	Strongly opposed	Moderately opposed
Rifle-Meeker	29.5%	27.3%	13.1%	30.1%
Delta	35.0%	39.0%	12.2%	13.8%

The attitudinal difference between the two groupings was emphasized by the fact that Meeker-Rifle respondents showed greater apprehension about the consequences of the project, while the Delta respondents expressed greater expectation of economic gain. Meeker-Rifle respondents had significantly more accurate knowledge about the proposed number of blasts than those in Delta.

The next step in the research involved an effort to determine which factors influenced the attitudes expressed about the project. Separate tests were run on the Meeker-Rifle and the Delta data. It was found that of all the variables in the Meeker-Rifle survey, apprehension was the best predictor of negative attitude toward the blast: the greater the apprehension, the less favorable the attitude toward the blasting. At the other end of the spectrum, anticipation of economic gain was found to be most significant: the greater the expectation of economic gain, the more favorable the attitude toward the blasting. As expected, these combined findings correspond to the separate results of the Meeker survey. But in contrast to the separate Meeker study, the combined Meeker-Rifle data demonstrated some correlation between age and sex and attitude toward the detonations. Older people were more likely to favor the blasting than younger people, and men were more likely to favor it than women. Knowledge, educational level, occupation, income, marital

146

status, and number of children were found to be of no significant predictive value in determining attitudes.

Analysis of the Delta data indicated that apprehension and expectation of economic gain were the strongest and only significant predictors of attitudes. As in the case of the Meeker-Rifle results, the greater the apprehension, the less favorable was the attitude of Delta residents. In addition, age, sex, occupation, and marital status emerged as indicators: older people, men, skilled workers, business men and professional people, and single and divorced persons were more likely to favor the blasting than younger people, women, unskilled or semiskilled workers, and married and widowed persons.

From these results, it seems clear that apprehension about the nuclear shots and expectation of economic gain are the best indications of attitude. As the Meeker-Rifle respondents indicated greater apprehension than did those of Delta, it follows that the Meeker-Rifle group were more opposed to the Rio Blanco Project than the Delta group. Similarly, Delta's expectation of economic gain to the general area of the White River Valley is consistent with its stronger support of the project. Apparently many of the people questioned in Delta believed that Project Rio Blanco would materially benefit the economy of the Western Slope. These results suggest that since their homes had not been, and would not be damaged by detonations, and since the people would not be directly exposed to possible increases in radioactivity, Delta residents more readily supported the project.

Even though the Rulison results indicated no increases in radioactivity following the blast, there is sufficient disagreement among scientists about the consequences of *any* increase in radioactivity to cause concern on the part of those who will receive such increased doses of radiation. Consequently, some Rifle residents were clearly concerned about further nuclear testing in their general area. Furthermore, the fact that large sums of money were paid out in damage claims following the Rulison shot, without the expected accrual of beneficial long-term effects on the economy of the area, seems to have dampened Rifle's enthusiasm for further testing.

In the light of such considerations, some Rifle residents probably felt they had more reasons for opposing than supporting a project similar to the Rulison experiment. Some Meeker residents as well might have shared such sentiments, but they had an additional reason for apprehension. In view of the publicly stated concern of oil-shale companies that the detonations might damage the rich oil-shale deposits of the Piceance Creek Basin, some probably felt that a less profitable natural resource was to be exploited at the expense of a potentially far more lucrative one. Though such considerations are at this time largely speculation, they do seem to offer a plausible explanation why the people who had experienced some phase or another of nuclear stimulation projects were more opposed to Project Rio Blanco than those who had had no such personal and direct experience.

Given the sometimes intense opinion on both sides of the issue, the question arises as to why so few local people made their feelings known at public meetings and hearings. Interviews conducted by members of the team showed that some people in the area were afraid they would make fools of themselves in public because of their limited education, while others felt that it was up to the authorities to make the right decision. Some local residents during 1972 felt that their views were being adequately expressed by the few vocal citizens who did make their views known.

Newspaper coverage undoubtedly had some effect in the formulation of citizen opinion on Rio Blanco. Local papers covered project planning as it developed, and expressed viewpoints both pro and con. The *Meeker Herald* generally confined itself to reporting project planning and citizen reaction to such plans. Reporting often included Chamber of Commerce meetings and therefore reflected the Chamber position, which was favorable to the project.[22] The *Rifle Telegram*, on the other hand, continued its outspoken opposition to any nuclear detonations, just as it had done in respect to Project Rulison.[23] The sponsors, as earlier noted, held periodic public meetings, featuring explanation of the project by various AEC and industry officials. In February 1973, they began printing and distributing the *Rio Blanco News* from Grand Junction. The purpose of the newsletter, according to the sponsors, was to "keep people informed." The first issue devoted itself to this purpose by announcing in large headlines that "Gas from Project Rio Blanco Unit Could Equal 10 Year Supply for the States."[24] Clearly, both proponents and opponents could find plenty of material to substantiate their views, and it was difficult for the average citizen to know what was truth and what was fiction.

148

It is also possible that in spite of intensely held views, the people in 1972 and early 1973 did not consider nuclear blasting as imminent. Although it had been publicized that a detonation would occur in 1972, no specific date was ever mentioned. It was publicly known that the concerned state agencies had not yet issued the necessary permits, and the Governor's position and role in the decision-making process was apparently not completely clear, even to the Governor himself. Furthermore, the AEC did not sign a final agreement with CER until 1973. In such a state of uncertainty, citizens having strong opinions on either side of the issue could feel that there was a good chance that their wishes as to whether or not the blast would occur would be realized. But as time went by, and it appeared that the Rio Blanco detonation would in fact occur in the spring of 1973, residents of the Rio Blanco area did turn out at meetings in increasing numbers, both to ask questions and to publicize their opposition.

A few weeks before the actual detonation, the sponsors held a series of meetings in Meeker, DeBeque, and Rifle, featuring Dr. John Toman, a project scientist from the AEC Lawrence Livermore Laboratory. Dr. Toman encountered both support—principally from those who felt Rio Blanco would help ease the gas shortage and improve the economy of the area—and hostility, as for example from those who wondered if the "AEC would explode one of those things in their own back yard."[25] At the Rifle meeting, the audience was particularly hostile, and Dr. Toman was continually interrupted by heated questions. In answering questions or statements about the dangers of the blast, Dr. Toman "continually countered [with the statement] that their information was hearsay, plain incorrect or based on an emotional fear of nuclear tests. . . ."[26]

It seems clear that Meeker residents were reasonably well informed about Project Rio Blanco and that many had formed strong opinions. It is therefore puzzling why there was relatively little public comment until just before the detonation. If local residents were not apathetic about the project, perhaps they felt that their individual opinions could have no substantive effect on the decision-making process, and therefore they remained silent. In any case, the outcome was that a nuclear detonation occurred with the acquiescence, if not necessarily the blessing, of Western Slope residents.

149

THE GOVERNOR, THE GOVERNMENT, AND THE PEOPLE

Western Slope acquiescence to Project Rio Blanco was ultimately shared by the Governor, though his attitude about nuclear detonations in the state seemed to be uncertain and ambivalent during various stages of project planning. There were some differences of opinion in several state agencies charged with responsibilities for the project, but in the end, they too acquiesced to the program. Some Colorado legislators supported the project; others opposed it. Finally, the issue of proceeding with detonation as planned was resolved by a combination of pressures which stemmed from an apparent conviction within the federal government that the completion of Phase I of the Project was in the national interest and should proceed as planned. The Court found no reason to disagree.

When the Governor announced in August 1972 that his Advisory Committee favored proceeding with Phase I of the Project, he noted that he was expecting to receive the results of a study by the Department of the Interior on the potential effects of the Rio Blanco blast. Later in the month, the Governor signed a federal-state-industry agreement to study the impact of oil-shale development on the Colorado environment. Two tracts of land near Meeker were to be covered by the study, including areas in Rio Blanco, Garfield and Mesa Counties which overlap the Rio Blanco nuclear stimulation site. The Governor thus was faced with potentially conflicting advice and was placed squarely in the middle of two aspiring, though not necessarily congenial, developers of his state's natural resources.

In March 1972, when the Governor's Advisory Committee had voted preliminary approval for Phase I of the Project, the Governor stated that he was undecided about permitting the project and that if the blast were allowed it would be "subject to fairly stringent regulation." He also said that approval for one blast would not imply consent for the whole program envisioned by the sponsors.[27]

In December 1972, the AEC and CER began negotiations over the actual contract governing the conduct of the experiment, and a tentative detonation date of March 10, 1973, was set. During a press conference on December 20, 1972, Governor Love made public a letter that he had written to James Schlesinger, the then new Chairman of the AEC. In this letter, he expressed misgivings about the compatability of oil shale development and the nuclear stimulation of gas in the same area. In advising Mr. Schlesinger about the probability of his veto of the project, Governor Love reminded him of his (Love's) earlier correspondence with Dr. Seaborg and his presumed right to halt nuclear detonations which he felt were not in the best interests of the state. He was writing, he said, to avoid needless further expense in time and money, should he decide to use his veto power.[28]

In making this letter public, the Governor launched a bombshell of his own, and reaction was not long in coming. Herbert Grier, the president of CER, was "dismayed" at this first indication of the Governor's position on the project. He announced that CER would take no action in cancelling plans for the project until the situation was clarified. Tom Ten Eyck, the State Director of Natural Resources, announced that a majority of the Governor's Advisory Committee did not share the Governor's view. An unusual combination of those representing oil shale interests and environmentalists made up the minority of the Commission who applauded the Governor's stand. The Committee vote in favor of the project was 12 to 5, but as Mr. Ten Eyck observed, "neither [the oil or gas] industry is home free, no matter what the Governor decides."[29] Reaction in Meeker was somewhat mixed, as the *Meeker Herald* noted in reporting that "several men seemed either indifferent to the shot or opposed it because they feared it would endanger development of an oil-shale industry."[30] An AEC spokesman meanwhile "reaffirmed the Governor's veto power over the project in saying, if he says it will be vetoed, it will be vetoed. It is that simple."[31]

Apparently, however, it was not in fact quite that simple. Chairman Schlesinger made efforts to set the matter straight in saying that while the AEC desired the concurrence of appropriate state officials, it was also recognized that the matter "is not a question of legal rights, but an example of intergovernmental comity. The federal government has the authority to proceed with such a test on federal lands." Mr. Schlesinger added that a "cancellation of the Rio Blanco test would reflect intergovernmental conciliation rather than the exercise of state authority." Schlesinger indicated his understanding of the Governor's concern over incompatibility of oil shale

151

development with nuclear stimulation of natural gas, but added that AEC personnel saw no such incompatability. In any event, he concluded that "the only way a firm resolution of this point will be obtained will be by conducting a test."[32]

With the continuation of the project in doubt, various local groups reacted quickly. The Colorado West Regional Council of Governments, a group of representatives elected from Garfield, Mesa, Moffat, and Rio Blanco Counties, adopted a resolution urging the Governor to approve the project. Although the appropriateness of taking such action without a prior polling of member communities was questioned, it was nevertheless decided that the project was directly concerned with "the livelihood of thousands of workers in western Colorado and [with] keeping people warm in winter," and the resolution was sent to the Governor.[33] The Rifle Chamber of Commerce a few days later voted 20 to 4 to draft a resolution urging the Governor not to veto the blast, though there was no discussion by the members as to why this stand was taken.[34] In early February of 1973, the Meeker Town Council voted to send a letter of support to the Governor in favor of "one nuclear shot."[35]

In the midst of conflicting opinion and uncertainty about his veto power, the Governor took a trip to Washington, D.C., and upon his return stated that members of President Nixon's White House staff were also studying the project. In a somewhat startling reinterpretation of his powers and those of the state in relation to federal authority, Governor Love said that he presumed that "the White House could make the decision if the President chooses."[36] It was later reported that the first indication that the test had been approved followed a meeting on February 6 with Secretary of Agriculture Earl Butz and White House aide Richard Fairbanks.[37] The AEC through its newly designated Chairperson, Dr. Dixie Lee Ray, left little doubt about the way the wind was blowing. Although declining to comment on the role of the White House, she said she was confident that the shot would go as scheduled for spring detonation and that neither the Department of the Interior nor Governor Love would oppose completion of Phase I.[38]

The Governor's assurances that Phase I would not jeopardize oil shale development or commit the federal government to further nuclear blasts in Colorado did little to appease opponents of the project. The fact that the federal government assumed legal liability for any residual radioactivity remaining underground was equally unimpressive to those who did not wish to have such radioactivity created in the first place.[39]

Governor Love's announcement drew sharply divided comment from opposite sides of the political fence. Representative James Johnson, the new Republican Congressman from Colorado's 4th District, which includes the Rio Blanco test site, said that he found no reasons to object to the test which could have beneficial results for the people in the area and which was in the national interest. Colorado's newly elected Senator Floyd Haskell, on the other hand, objected strenuously. While he conceded that one shot might not be harmful, the case would be far different if 140 to 500 shots were used in full field development. Such a series of shots, in his view, had the potential for releasing "immense amounts of radioactivity from plutonium sources that have a life expectancy into the millions of years."[40] This, Senator Haskell concluded, was a risk we could not afford. Since he felt that full field development would be unacceptable under any conditions, he strongly questioned the wisdom of going ahead with even a single nuclear blast.

In the face of decisions apparently already taken, it seemed somewhat anticlimatic, if not futile, to proceed with hearings by two state agencies which had, at least in theory, some authority over the project. Nevertheless, the chairmen of both the Oil and Gas Conservation Commission and the Water Pollution Control Commission said they would press on with inquiries about the project on the basis of their statutory Colorado authority. The Colorado Water Pollution Control Commission has a broad statutory authority to inquire into public need for any project which might adversely affect either ground or surface water in the state. A permit is required from the Colorado Oil and Gas Conservation Commission before any oil or gas may be extracted from lands in the state. The Governor, when asked if he would support a veto of the project by either Commission, was somewhat equivocal in saying that it would depend upon the grounds the Commission used in reaching its decision.[41]

Another state agency, the Colorado Board of Health, also was clearly interested in Rio Blanco, but was lacking in specific statutory authority to exert control over it. In an effort to gain early widespread support for the project, CER had requested, through its Denver attorney, that the Board of Health hold public hearings.[41] The Board was determined to exert as much authority as possible to prevent a recurrence of the jurisdictional gap which existed between the Health Department and the AEC in respect to the use of uranium mill tailings in Grand Junction, Colorado, and the alleged plutonium contamination surrounding the AEC-Dow Chemical Company Rocky Flats plutonium recovery installation near Boulder.

In these instances, contractors dealing with nuclear materials within the state had not been licensed by the state, with the result that the Board of Health lacked any basis for control of their activities. In the words of Chairman Glen E. Keller, Jr., the Board was "left holding the bag," charged with "ultimate responsibility" for the health and welfare of the people of the state—but lacking the original jurisdiction necessary to accomplish the task.[43]

The history of jurisdictional difficulties that the Public Health Department had with the AEC over the control and ultimate responsibility for radioactivity which escaped in Grand Junction and at Rocky Flats made the Board especially wary of some parts of CER's testimony at the public hearings that were held on the project. One such clause explained that "the Atomic Energy Commission will retain exclusive jurisdiction over all gaseous radioactive material until such time as CER . . . might ultimately obtain a license from the AEC or the State of Colorado as appropriate."[44] As Mr. Keller saw it, this statement was similar to a "red flag forecasting the possibility that there might be an effort on the part of our favorite uncle to preempt the state under the commerce clause or some other legal device from regulating the amount of radiation the state would allow to be present in the [Rio Blanco] gas." Although CER readily agreed to cooperate with state licensing procedures, there was no firm agreement during the course of the hearings as to precisely what radiation standards would be used. Nor was there any precise delineation of the state's authority to require conformity to such standards as might be adopted.

The first hearing before the Colorado Board of Health was held in February 1973; it involved four hours of testimony from six representatives of CER, as well as critical comment from such people as Dr. H. Peter Metzger, David Evans, Charles Gaylord, and others. Mr. David Cullen, general counsel for CER, made the point that the AEC would retain control over all phases of the project, from providing three nuclear devices for detonation to disposing of waste radioactive substances and protecting public health and safety. Damage compensation would be provided by CER for local claims of up to $100,000, by an additional $11 million in insurance by private companies, and beyond that, by U.S. Government indemnity up to a liability of $560 million.

Additional testimony from CER officials dealt with estimates of the amount of natural gas in place in the Rio Blanco field, with the eventual economic feasibiity of nuclear stimulation with multiple detonations, and with engineering aspects of the project. The latter included plugging of the drill hole, wellhead capping, and the installation of 6 test wells and 29 seismic stations to monitor the effects of detonation. Tentative plans for the disposal of radioactive water produced by the nuclear detonation were mentioned—reinjection of the water into a well at the depth of the original nuclear blast. The company Director of Health and Safety described safety precautions which were based on "a maximum credible accident possible," including an assembly center for the 400 area residents, the evacuation of a deer herd, traffic routing precautions, and evacuation of mines. Further testimony from CER's radiation consultant indicated that potential radiation dosages in water and air would be well within AEC's millirem standards anywhere within a distance of 25 to 30 miles from surface ground zero, for any conceivable well seepage and plume or wind patterns for a period of 35 days. Testimony also was given on the predicted extent of the expected seismic shock and on a technique of holding property damage to a minimum by pre-shot bracing of vulnerable structures.

After the formal CER presentation, David Evans, an opposition witness again representing the Colorado Open Space Council and the Denver Audubon Society, termed the CER proposals "dangerous and unacceptable." He based this conclusion on the fact that the ultimate objective of full field development would involve drilling and nuclear stimulation of 140 to 280 wells, each containing 3 to 5 nuclear devices.[45] Eventually, projecting nuclear stimulation of natural gas on a national scale, up to 13,000 wells would have to be stimulated. This would require 20 blasts of 30 to 100 kiloton bombs per week for a period of 12 years. Such a program, Evans said, would deposit hundreds of millions of curies of long-lived radioactivity underground, including strontium-90 and cesium-137 in water-soluble form. He asked, very basically, who would inspect the wells 50 years after gas production has ended, and which agency would be responsible for checking and repairing corroded well casings for 400 years to stop leaks of poisonous radioactivity? Evans also warned that once long-term radioactive material entered the Colorado River system, the damage would be done and the pollution irreversible.

Both Mr. Evans and H. Peter Metzger questioned the economic rationale of the sponsors. Evans also expressed the opinion that hydraulic fracturing techniques could be used economically at the prevailing commercial price of 20 to 30 cents per thousand cubic feet. Dr. Metzger noted that all the commercial firms involved in nuclear stimulation experiments, including CER, El Paso Natural Gas Company, Austral Oil Company, and the Colorado Interstate Gas Corporation, charge off wildcat or new venture losses in their rate structures. Because of the fact that company capital isn't invested initially in the risks of a nuclear stimulation experiment, a large part of the bill is paid by the public. Further adverse testimony came from Dr. Charles Gaylord, Chief of Radiology at National Jewish Hospital in Denver. Dr. Gaylord expressed concern about the past record of the AEC in supervising radioactive waste disposal in the state of Colorado. In his words, "Whenever [the AEC] comes into the state, the problem ends up in this room," or the State Board of Health's hearing room. His general contention was that too little is known about the long-term effects of exposure to radiation to proceed with projects that could greatly increase the radiation level. In his view, the potential danger to public health and welfare inherent in the project was too great a risk to take with current knowledge about radiation effects.

After completion of this testimony, the Board decided to continue the Rio Blanco hearings at a later meeting, which would follow scheduled meetings of the Colorado Water Pollution Control Commission and the Oil and Gas Conservation Commission. Presumably, these two agencies, which have statutory jurisdiction over the project, were expected to make decisions relevant to public health matters and thereby provide some guidance to the Board.

The Colorado Oil and Gas Conservation Commission met in a marathon session on February 20, 1973, and a panel of CER participants again testified to both the need for, and the planned operations of, the project. Representatives of TOSCO and International Resources, Inc., which owns 8,300 acres of salt leaseholds 7 miles from the project test site, were on hand to present testimony and question the sponsors.[46] A number of the eight Oil and Gas Conservation Commission members present at this meeting acknowledged possible personal conflicts of interest because of their oil, oil shale, or natural gas involvements. Commissioner Godding stated that he held stock in the Equity Oil Company and is a client of a Denver firm retained by CER Geonuclear Corporation, Commissioner Dunn acknowledged that he had performed land acquisition work for Equity Oil and had oil and gas interests in the Piceance Creek Basin area, and Commissioner Juhan admitted that he owned Equity Oil Company stock and had business connections with TOSCO. None of these potential conflicts of interest, however, were deemed serious enough to warrant disqualification of individual members during the Rio Blanco hearings.

In a surprise move, CER officials withdrew their earlier original request for a permit for a unit operating agreement for compulsory use of about 95,000 acres in the Piceance Creek Basin. In effect, this move limited the scope of the hearing to the granting or withholding of permission for a single shot known as RB-01. If CER had not taken this position, the Commission would have been in the position of ruling on the sponsors' plan to proceed ultimately with full field development and the detonation of a large number of nuclear devices. TOSCO representatives were thus preempted from giving their prepared testimony in opposition to full field development. However, the President of Industrial Resources, Inc. testified that mineral deposits in his company's leaseholds were solidly intact and could be damaged by even one nuclear detonation. He presented his company's view that CER should be requested to monitor the sodium salt deposits in both pre- and post-shot conditions and should be prepared to pay for any blast-related damage to their mineral deposits.

157

As a result of fears often mentioned at the hearings, CER lawyers again stated that the AEC-CER contract, at this time presumably near completion, would contain a provision requiring prior approval by the State of Colorado before the company would accept custody or control over any nuclear materials. The lawyers continually stressed the fact that the hearing concerned only the first Rio Blanco test shot and that further application to the Commission would be made for an operating agreement, including approval of a water-disposal well, when and if that should become necessary.

A great deal of CER testimony before the Oil and Gas Commission was technical in nature, with the same representatives covering much the same sort of material as was presented at the Public Health hearing. As a result of cross-examination by TOSCO personnel, however, particular emphasis was placed on the possibility of blast-created spall or shock waves affecting the oil-shale-bearing formations. CER specialists produced data in support of their prediction that the maximum spall effect would not damage oil-shale deposits. The Lawrence Livermore Laboratory calculations, which were used as a basis for CER predictions, projected a maximum spall depth of 400 feet, while TOSCO calculations indicated a spall depth of 1,200 feet. These figures, of course, are only predicted estimates of the depths to which shock waves would reach down to lower geological layers.

In these and other respects, there was no meeting of the expert minds. David Evans, who once again testified, stated that he was no more convinced of the need for the project than he had been at the earlier hearing. At the conclusion of the meeting, Chairman Milton Hoffman said that disapproval of the project was "a definite possibility."[47]

The following day (Feb. 21, 1973), the Colorado Water Pollution Control Commission heard from several CER witnesses who testified to the necessity and safety of the project.[48] The thrust of the presentation centered on the contention that there would be no ground-water pollution resulting from the blast, although the question of what would happen to the estimated 272,000 gallons of radioactive water produced by the nuclear explosion was not specifically answered since CER made it clear that it was not seeking a water-disposal permit at this time. Instead, CER promised that water-disposal plans would be presented to the Commission at an appropriate time, after they had been clearly formulated.[49]

The Water Pollution Control Commission took no position at the hearing on whether Project Rio Blanco was in the public interest or whether water pollution would result from it. Instead, it was decided that further written testimony would be received until March 12, and that the full Commission would consider all testimony at its March 20th meeting before taking any official action.

POLITICS AND THE BUREAUCRACY: POLITICAL REVERBERATIONS

Shortly after the various state commission hearings, Senator Floyd Haskell announced that he had formally asked the Department of the Interior to suspend the Rio Blanco Project. On the basis of what he termed an exhaustive staff study and consultation with several scientists, he concluded that full field development involving up to 500 nuclear detonations is "absolutely out of the question."[50] At a press conference on March 2, the Senator released copies of letters he had sent to Rogers C. B. Morton, Secretary of the Interior, AEC Chairperson Dr. Dixie Lee Ray, and Governor Love, in which he asked that the test be suspended for two basic reasons:

> First, if full field development is out of the question— and there seems to be unanimity in this regard—why go ahead with a single test that has possible danger?

> My second point . . . is based upon the National Environmental Policy Act. This shot is a part of a project. I have been advised that under Section 102 (2) (c) of NEPA that an environmental impact statement should be made on the project as a whole, and this has not been done. Therefore, unless and until this is made it would not be legal to go ahead with the shot.[51]

The Senator also said that he had written to Dr. Ray on February 14, 1973, asking for detailed information on the quantity of each radioactive element that would result from the shot and for the Commission's recommendations as to disposal or containment of these materials. Although he had requested a speedy reply, he had not yet received an answer. He said that should his efforts to stop the shot fail, he would initiate legislation to bar any further funds for the project. In addition, he said there was a possibility that legal action would be taken to halt project operations. The Senator continued his own efforts to achieve this objective by telephoning Governor Love twice and then returning to Colorado for a personal interview. Although he failed to convince the Governor that the project should be stopped, both men apparently agreed that full field development was out of the question.[52]

On May 3, the AEC announced that an "addendum" to the final Rio Blanco environmental impact statement was being prepared, including "late material" which was not available when the final statement had gone to press.[53] According to the *Meeker Herald* and the *Rifle Telegram*, both AEC and Interior Department officials were attempting to ward off the threat of a law suit by the National Resources Defense Council because the AEC had not answered questions put to it about its April 1973 Rio Blanco Environmental Impact Statement. It is possible that the addendum was issued as an attempt to satisfy the complainants. In a statement released with the addendum, it was noted that application had been made for formal withdrawal of 360 acres of public land at the test site from the control of the Bureau of Land Management (Dept. of the Interior) to the authority of the AEC. Since the transfer had not been completed and the Secretary of the Interior had not issued a specific permit for any nuclear shot, the test would be delayed from March until May of 1973.

There apparently had been a continuing confusion of authority over these acres which came to light only after a BLM field survey of the test site in late January. It was discovered at that time that power transmission lines were being constructed at the project site without a required BLM permit. Construction was halted when the area BLM manager at Meeker discovered that earth had been dug up and piled three and four feet high, trees uprooted, and drainage washes filled without provision being made for water runoff. He concluded that "disturbances are taking place without any constraints, environmental regard, or authorization by the Bureau of Land Management of U.S. Geological Survey as far as we can ascertain."[54] An official of CER responded that environmental and erosion problems were not severe and that the construction in question was carried out under provisions of oil

160

and gas leases held by the project's co-sponsor, Equity Oil Company. CER had initially applied for BLM special use permits, but since they had not been granted, the companies decided they could proceed under the terms of the Equity leases. Ultimately, the transfer of authority to the AEC negated any need for BLM permits, but in the meantime, CER moved some of its construction installations to private land to minimize what was termed "federal red tape."

What was irritating red tape to project sponsors was welcome delay to environmental critics who urged indefinite postponement or cancellation of the project. Consumer advocate Ralph Nader, who spoke to a large gathering in Denver in April 1973, said that through Project Rio Blanco, the AEC was using Colorado as a "guinea pig" to get a foot in the door for as many as 1,500 nuclear explosions. What was needed, he said, was immediate legal action to postpone the detonation and the establishment of forums in Washington, D.C., and Colorado to meaningfully inform the public and their elected leaders about the real stakes involved in nuclear experimentation.[55]

Meanwhile, at an Earth Week program at Colorado State University in Fort Collins, Dr. Carlos Stern, a professor of environmental economics at the University of Connecticut, and a former energy consultant at the Environmental Protection Agency, presented some harsh conclusions of his own about Project Rio Blanco. A study done with Dr. Emma Verdieck, also of the University of Connecticut, led Stern to conclude that the AEC and allied companies should stop "selling the program and explain the uncertainties." Among these uncertainties, he cited "the industry's responsibility for damage claims, the risk of earthquakes if many shots are undertaken, the safe disposition of the first year's gas production which would be highly radioactive, and the economic justification of the program." Stern suggested further that the AEC's domination of research funding for nuclear projects precluded emphasis on non-nuclear alternatives, which might be economically competitive and at the same time avoid the nuclear risks inherent in the Plowshare Program.[56]

The Democratic side of the aisle in the Colorado House of Representatives was apparently stirred by anti-Rio Blanco opinion and analysis. A minority coalition of 26 House Democrats signed a statement that they felt the Rio Blanco

161

detonation would make little sense without further shots to follow it up. In their view, further investigation of non-nuclear conventional technologies should be explored before carrying out the Rio Blanco test. The AEC, they said, should accept "the burden of proof in justifying Project Rio Blanco if environmental considerations suggest no further field development."[57] State Representative Richard Lamm, a Democrat from Denver and an active environmentalist, went further in urging Interior Secretary Morton to halt the project. At a Colorado Public Health Association seminar, Lamm said he thought the Governor would not stop the explosions "even though he clearly has the authority to do so."[58] He, along with Senator Haskell, questioned the legality of AEC procedures and their conformity with provisions of the National Environmental Policy Act.

The AEC, meanwhile, was continuing negotiations with CER, resulting on April 13 in a formal contract to carry out the project. CER, as the principal contractor, was to pay 85 percent of the estimated $7.5 million cost. Two days later, the special permits allowing completion of communication facilities and power installations at the test site were signed by the Colorado Director of the Bureau of Land Management, Mr. Dale Andrus.[59] Senator Haskell promptly announced that he would hold hearings on Project Rio Blanco on May 11 in his official position as a member of the Senate Interior Subcommittee on Public Lands. The AEC, he said, had been asked for information about the nuclear device that was to be used and had "volunteered to give me a classified briefing." The Senator made plain his view that the project "affects the health and welfare of the citizens of Colorado and should not be handled behind our backs. The AEC must be able to justify what they are doing in open hearing." He also said that he was continuing to press the Department of the Interior for a response to his question as to "why conventional drilling practices have not been tried prior to endangering our welfare with radioactive tests."[60]

Nonetheless, the Colorado Oil and Gas Commission completed its deliberations by giving formal permission for detonation on April 27, but added the stipulation that any further nuclear shots beyond those scheduled for Phase I would require additional hearings and permits. A further restriction required the sponsors to notify the Commission at least 60 days before well re-entry and testing took place. On the same day, the Acting Secretary of the

Interior signed an order transferring the control of the 360 acres of public lands at the test site from BLM control to the AEC. With this last remaining federal action completed, the only thing standing in the way of a mid-May detonation was the State Water Pollution Control Commission.

That commission voted unanimously on April 17, 1973, to issue a permit for the detonation. They declined to hear any further testimony, basing their decision on the material that was presented at their February hearing. Acting Chairman Howard Lewis said that the Commission would consider the problem of disposing of the radioactive water produced by the blast after a specific proposal for its disposal was presented.

The State Board of Health, for its part, expressed "most serious concern" over Project Rio Blanco, but acknowledged in a 5 to 1 favorable vote that in fact it had no statutory control which would permit it to prevent detonation. A part of the statement issued by the board is indicative of the basis for its concern: "Mankind generally has insufficient experience with man-made radiation to be able to predict with any reasonable certainty the effect upon generations to come of any exposure in addition to that normally existent . . . . The wisdom, from the point of view of public health and safety, of creating further deposits of man-made radiation within the state has not been demonstrated to this Board."[61] Mr. Glen Keller, the Board's chairman, expressed his dissatisfaction that CER officials had not provided a satisfactory contingency plan for the protection of public health and safety in the event of a maximum possible accident. He felt, however, that the Health Department staff plan would probably be adequate to the need for such protection. The lone dissenting vote against the project came from board member Mary Alice Munger, who said that if Phase I of the project was permitted, the AEC through a process of gradualism and "one shot at a time" would ultimately gain the objective of full field development. Even one shot, from her point of view, was one shot too many.

THE TEMPO INCREASES

For reasons still unexplained, the AEC suddenly decided to speed up the timetable for emplacement of the nuclear devices that were

163

to be used in the project. Emplacement, originally scheduled for May 9, was moved up to May 3, but in fact this date was not adhered to because of adverse weather conditions. This unexpected change in plans brought a sharp response from Senator Haskell, who still planned a May 11 hearing on the project in the Senate Interior Subcommittee: "If the hearing results in new evidence as I believe it will, there will be no way to stop the device from being buried. Neither the Atomic Energy Commission nor . . . CER Geonuclear has made any plans for removing the device[s] from the ground should there be a reason not to continue."[62] In a speech on the Senate floor, he charged that the AEC, backed by the administration, was ignoring Congress, which he contended was "yet another encroachment upon the prerogatives of the institution."[63]

In the hearings, which took place after the three nuclear devices had already been cemented in place, Senator Haskell still had hopes that detonation could be prevented. He questioned Gerald W. Johnson, Director of the AEC Applied Technology Division, as to whether the AEC would feel that a program to develop 5,680 wells with more than 22,000 atomic explosions would be desirable. Director Johnson replied that "we've not been asked that question." When informed by Haskell that he was being asked that question now, Johnson replied that the function of the AEC was only to provide technology, whereas responsibility for long-range decisions rested with other agencies of government.

A number of other people testified at the hearings taking positions both for and against the project. A Rio Blanco county commissioner expressed exasperation that some of Colorado's elected representatives seemed to pay less attention to the local area residents who favored the project than to some "so-called experts who live as far away as Connecticut." The former president of the American Geological Society, Dr. Richard Morse, took an opposite position, saying that he would bet "10 to 1 that hydraulic fracturing would be successful in the Piceance Basin." The entire Wyoming Congressional delegation and newly elected Colorado Representative Pat Schroeder, a Democrat from Denver, testified against the project. Both Representative Schroeder and Senator Gale McGee, a Democrat from Wyoming, felt that the AEC had withheld pertinent information. Mrs. Schroeder said that "in order to prevent further Rio Blancos we must demand, by appropriate legislation if necessary, candor from the AEC."[64]

164

Glen Keller of the Colorado Board of Health reiterated concerns that he and the Board had already expressed, and he added that he was surprised at the willingness of the AEC and CER to proceed with Rio Blanco without "first determining in a public forum the safety of the gas to be produced." Though Senator Haskell was able to get the AEC to declassify a list of secondary radioactive substances which would be produced by the blast, he was unable to achieve his goal of preventing the detonation.

During the month of May, a number of groups opposed to the test took various actions in an attempt to postpone or cancel the shot. The Arapahoe Medical Society, following the lead of the 2,600 member Colorado Medical Society, resolved that the medical impact of the project required further evaluation, and declared that "potential risks exceed the reasonably anticipated gains."[65] Thirty-one professors of science at Colorado State University and eleven professors from the University of Denver School of Law faculty joined in sending a letter to Senator Peter Dominick, Republican of Colorado, urging his intercession to stop the project. (Senator Dominick, though a member of the Joint Committee on Atomic Energy and a vigorous advocate of both Gasbuggy and Rulison, did not make any public comment on Rio Blanco.)[66] In a last-ditch, but equally unsuccessful attempt, eleven conservation groups appealed in a joint letter to President Nixon to halt the blast. As no action was taken on these requests in Washington or elsewhere, it appeared that the only hope for critics of the project lay in a lawsuit filed by Citizens for Colorado's Future and other interested persons.

History thus seemed to repeat itself. As it came down to the last days before scheduled detonation, the action in both the Rulison and Rio Blanco stores was much the same. It was the court, in each instance, that was summoned to resolve the problem of allowing or disallowing underground nuclear experimentation. A judge was once again faced with conflicting testimony and opinion from the scientific community and basic disagreement among citizens as to the wisdom of pursuing the nuclear route to ease the national energy crisis. The analysis of the technical side of Rio Blanco in Chapter 6 illustrates the kind of scientific and engineering complexities involved in the project which the court was once again called upon to evaluate.

NOTES

[1]A referred proposal may be placed on the Colorado ballot through citizen petition. The number of valid petition signatures must equal 10 percent of the number of votes cast in the last general election for the office of Secretary of State.

[2]*Denver Post*, July 20, 1972.

[3]Copies of correspondence from Governor Love's office, State Capital Building, Denver, Colorado, July 1972.

[4]*Denver Post*, April 19, 1972.

[5]*Calvert Cliffs Coordinating Committee v. U. S. Atomic Energy Commission, Federal Reporter 2d* (499 F. 2nd), 1109, *Environmental Reporter Cases* (2 ERC) 1779, *Environmental Law Reporter* (1 ELR) 20346 (D.C. Circuit Ct., 1971).

[6]*United States Code Annotated* (42 USCA Sect.), 4881.

[7]Appendix D. 10 *Code of Federal Regulations* (CFR), Sect. 20.

[8]*Calvert Cliffs Coordinating Committee, Inc., et al., plaintiffs, v. U.S. Atomic Energy Commission*, nos. 24,889 and 24,871, July 23, 1971, ERC 1770.

[9]Nos. 24,839 and 24,871, July 28, 1971, 2 ERC 1779.

[10]2 ERC 1788.

[11]36 *Federal Register* (FR) 18071.

[12]36 FR 18072.

[13]This restatement is taken from the *Meeker Herald*, March 2, 1972, the local newpaper in the town where one of the hearings would be held.

[14]USAEC Transcript, *Informal Public Hearings on Project Rio Blanco*, Meeker, Colorado, March 24, 1972, p. 15.

[15]*Ibid.*, p. 85.

[16]*Meeker Herald*, Jan. 20, 1972.

[17]*Meeker Herald*, Feb. 17, 1972.

[18]At the point that the survey was begun it appeared that the project would be implemented during the spring of 1972, rather than a year later as in fact was the case. It is, of course, possible that there may have been changes in attitude between the time of the survey and actual detonation. The survey was conducted by faculty, students and staff of the Technology, Law and Politics Research Group at the University of Colorado.

[19]There are some limitations in using a phone book from which to draw a sample, but it is generally felt that the method provides a reliable listing of the total population in a small-to-medium size town. See D. A. Leuthold and R. Scheele, "Patterns of Bias in Samples Based on Telephone Directories," *Public Opinion Quarterly*, Summer 1971.

[20]These characteristics included size, population growth from 1960 to 1970, predominant occupations, age and sex of residents, and types of housing. Eastern Slope towns were eliminated because those with population sizes similar to Meeker are in tourist areas; towns on the plains were not selected on the theory that most residents would not be at all familiar with the project; towns

in southern Colorado were not selected because of the large migrant worker population in those areas; these are not found in Meeker.

21The questions were the same as those sent out to Meeker, except that the names of the local newspapers were changed.

22*Meeker Herald*, March 30, 1972; April 20, 1972.

23*Rifle Telegram*, March 23, 1972; April 6, 1972.

24*Rio Blanco News*, Vol. 1, Feb. 1973.

25*Boulder Daily Camera*, May 7, 1973.

26*Boulder Daily Camera*, April 25, 1973.

27*Rifle Telegram*, Aug. 2, 1972.

28Reports of press conference, *Denver Post*, Dec. 21, 1972, and *Boulder Daily Camera*, Dec. 21, 1972.

29*Denver Post*, Dec. 23, 1972.

30*Meeker Herald*, Dec. 28, 1972.

31*Denver Post*, Dec. 26, 1972.

32Copy of letter from James R. Schlesinger to John A. Love, Jan. 11, 1973. See also "AEC Attorney 'Clarifies' Policy on State Plowshare Opposition," *Grand Junction Sentinel*, March 19, 1973. For a discussion of the Governor's loss of veto power from the viewpoint of a member of the press, see Sue O'Brien, "A Veto Lost," in "The Unsatisfied Man," *The Review of Colorado Journalism*, Denver, Colorado, March 1973.

33*Meeker Herald*, Jan. 11, 1973.

34*Rifle Telegram*, Jan. 17, 1973.

35*Meeker Herald*, Jan. 17, 1973.

36*Boulder Daily Camera*, Feb. 3, 1973.

37*Rocky Mountain News*, Feb. 9, 1973.

38*Meeker Herald*, Feb. 6, 1973.

39*Rocky Mountain News*, Feb. 9, 1973. The Governor's decision was announced after he had received letters from both the Secretary of the Department of the Interior and from the Chairperson of the AEC. For an analysis of the sequence of Governor Love's actions on Project Rio Blanco, see Robert Threlkeld, "Rio Blanco Blast to Go As Scheduled," *Rocky Mountain News*, Feb. 9, 1973.

40*Denver Post*, Feb. 10, 1973.

41*Rocky Mountain News*, Feb. 21, 1973.

42*Rifle Telegram*, Jan. 24, 1973.

43*Rocky Mountain News*, May 3, 1973.

44*Denver Post*, April 5, 1973. The following information is based on hearings held by the Board of the State Department of Public Health on Feb. 14, 1973, March 21, 1973, April 4, 1973, and May 3, 1973. See also *Rifle Telegram*, Jan. 24, 1973, *Denver Post*, Feb. 2, 1973, and March 22, 1973, and *Rocky Mountain News*, May 3, 1973.

45By Mr. Evans' estimate, a minimum of 420 and a maximum of 1,400 nuclear explosives would be entailed for full field, economically profitable development. This analysis of economic profitability, however, does not include the original costs of the nuclear devices themselves. Hearings before the Colorado Oil and Gas Conservation Commission, Feb. 20, 1973.

[46]Hearings before the Colorado Oil and Gas Conservation Commission, Feb. 20, 1973. See *Rocky Mountain News*, Feb. 21, 1973.

[47]*Rocky Mountain News*, Feb. 21, 1973.

[48]Colorado Water Pollution Control Commission Hearings, Feb. 21, 1973. See also the *Denver Post*, Feb. 22, 1973, and *Rocky Mountain News*, Feb. 22, 1973.

[49]See Chapter 7 for the radioactive water disposal difficulties which CER and the Water Pollution Control Commission have subsequently encountered.

[50]The material on Senator Haskell's part in the Rio Blanco proceedings is taken from staff reports on the project provided by Senator Haskell's office. See also the *Denver Post*, March 2, 3, 1973.

[51]Statement by Senator Haskell, March 2, 1973.

[52]*Denver Post*, March 4, 1973.

[53]*Meeker Herald*, March 13, 1973.

[54]*Denver Post*, March 11, 1973.

[55]*Denver Post*, April 10, 1973.

[56]*Denver Post*, March 18, 1973.

[57]*Ibid.*

[58]*Boulder Daily Camera*, March 18, 1973. Mr. Lamm, elected Governor of the State in 1974, may now have an opportunity to test his perception of the Governor's authority.

[59]*Denver Post*, April 15, 1973.

[60]*Boulder Daily Camera*, April 15, 1973.

[61]*Rocky Mountain News*, May 3, 1973.

[62]*Denver Post*, May 2, 1973.

[63]*Ibid.*

[64]*Rocky Mountain News*, May 12, 1973. Wyoming residents and their Representative opposed to the nuclear stimulation project, Wagon Wheel, in their state were at least temporarily more successful in stopping proposed plans than were their Colorado counterparts. U.S. Representative Teno Roncalio, Democrat of Wyoming, testified before the Public Works Subcommittee of the House Appropriations Committee to urge that a Nixon administration request for a $3.8 million appropriation to the AEC for underground nuclear explosions be eliminated. Shortly thereafter, Dr. Dixie Lee Ray announced that Project Wagon Wheel was dead for the foreseeable future. See the *Denver Post*, May 22, 1973.

[65]*Rocky Mountain News*, May 1, 1973.

[66]*Denver Post*, May 6, 1973. Senator Dominick's position on natural gas production has been to favor deregulation of the price of natural gas at the wellhead to provide additional funds for increased exploration of new reserves and resources. In the 1974 general election, Senator Dominick was defeated in his bid for reelection by Democrat Gary Hart of Denver.

Chapter 6

Technology and Its Evaluation: What Price Rio Blanco?

Both commercial sponsors of the Rio Blanco Project had previous experience in the development of energy resources in the Rocky Mountain West. CER Geonuclear Corporation has had long experience in nuclear projects and, as noted in Chapter 5, was project manager of the Rulison test. The Equity Oil Company, based in Salt Lake City, Utah, has been involved with oil-shale development in the Piceance Creek Basin since 1948, and with explorations for natural gas in that same area since 1954. Data which were collected in the process of drilling more than 50 conventional natural gas wells led the company to conclude that large natural gas reserves existed in the Piceance Creek Basin, but that commercial recovery of most of the gas would be uneconomical by conventional methods. Equity decided, therefore, to retain CER in September 1969 to study the feasibility of using nuclear stimulation recovery methods.[1] At this point, the groundwork was laid for the next, and to date, the last, underground nuclear explosion for peaceful purposes in the United States.

An initial feasibility study indicated that it would be possible to develop the entire Equity field, but that further, more detailed investigations should be made before a firm decision could be reached. To this end, Equity and CER signed a joint venture agreement on March 24, 1970. CER was to bear the cost of further testing in the Rio Blanco field, as well as the entire private cost of Phase I of the Rio Blanco experiment, which was estimated to be $3,500,000.[2] Upon performance of the agreement, CER would gain a 50 percent interest in the Piceance Creek Basin leases held by Equity.[3]

As the next step, CER sent a formal proposal, along with the feasibility study, to the Atomic Energy Commission for consideration as the next project in the Plowshare Program. On December 18, 1970, CER and the AEC signed a Project Definition contract, authorizing further study of technical feasibility and safety. One month later, the AEC assigned to the Lawrence Livermore Laboratory the task of supplying the nuclear explosives and associated services which would be required to carry out the project. The AEC Nevada Operations Office was designated as the Project Contracting and Operations Office.[4]

The specific area for the experiment was designated as the Rio Blanco Federal Oil and Gas Unit by the U.S. Geological Survey on September 30, 1971. The land area of this unit covers a total of 93,762 acres, with Equity and CER holding leases on 50,473 acres or 54.27 percent of the entire unit.[5] Figure 6-1 shows the total unit in relation to the Piceance Creek-Yellow Creek drainage basins. Although it was considered possible to subject the entire drainage basin to nuclear stimulation, the original proposal called for development only within the outlined area.

THE PROJECTED PLAN

In the Project Demonstration Plan, CER made the following preliminary analysis:

> Based on geological and geophysical log analysis of wells in the area [the 93,762 acre area, not the entire basin], the gas in place in this area is estimated to be 10- to 12-trillion standard cubic feet in the Fort Union Formation and the upper 1,200 feet of the Mesaverde Formation.

If all of the acreage in the unit area proved commercial, it is estimated that 35% of the gas in place, or 3.5 to 4.2 trillion standard cubic feet, can be recovered in the first twenty-five years assuming one stimulated well per section [640 acres]. If the field is developed with more than one well per section or if the gas is produced for more than twenty-five years, the percentage of gas and total recovery would be substantially increased.[6]

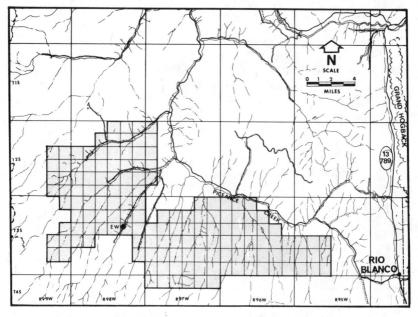

Figure 6-1. Project Rio Blanco experimental and development area location within the Piceance Creek-Yellow Creek drainage basin. (CER Geonuclear *Project Rio Blanco Definition Plan, Vol. I. Project Description.* Feb. 14, 1972, p. 8.)

The development plans provided for four distinct phases of operation with each phase dependent upon the successful completion of the prior one. Phase I was to consist of the simultaneous detonation of three nuclear explosive devices in a single well. It was originally scheduled for detonation in early 1972 but was ultimately delayed until May 1973. Figure 6-2 shows the location of the Phase I test site in relation to surrounding population centers.

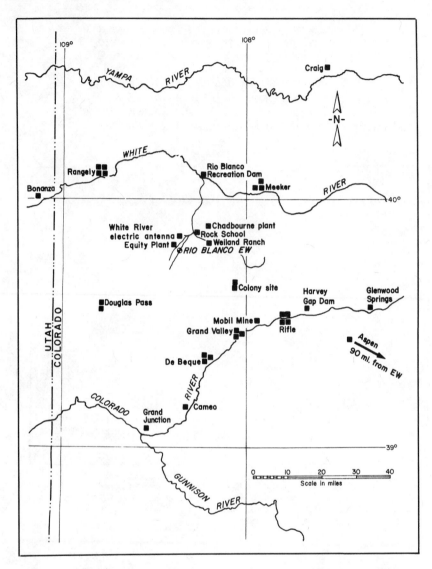

Figure 6-2. Project Rio Blanco site map. (C. E. Williams and G. R. Luctkchans, *Project Directors' Completion Report D + 30 Days,* NUO-135, July 1973, p. 19.)

172

Phase II of the plans projected the development of four to six wells, each stimulated by three to five nuclear explosive devices, or a total of twelve to thirty explosives. The detonation of the devices in all the wells on the same day was considered possible. In each of the wells in this phase it was projected that between 1,500 to 2,500 vertical feet of formations containing natural-gas-bearing rock would be stimulated. CER planners felt that Phases I and II would probably be decisive in determining the future of the entire technical program since the data provided by these two phases would enable industry and government to decide whether to make the substantial investments necessary for full field development. Phase II was to be integral to the entire plan in that it would supply additional data on nuclear engineering, the sequential firing of nuclear devices, and optimum field development techniques. These data, it was hoped, would contribute to the development of guidelines for the sale and use of natural gas containing trace amounts of radionuclides, the development of machinery to produce the required nuclear explosives, and the initiation of administrative procedures for the effective management of a large number of nuclear stimulations per year.[7]

Mr. Paul Dugan of the Equity Oil Company did not expect either Phase I or II of the project to show a profit because these tests would still be primarily experimental in nature. Phase I was designed as a test to determine whether nuclear stimulation techniques would produce enough gas to provide a reasonable financial rate-of-return. If, however, such a return were not achieved, the remainder of the plans for the project were to be scuttled. Assuming that Phase I would be successful, six Phase II shots would be carried out in scattered locations throughout the Piceance Creek Basin to determine the areas most profitable for further nuclear stimulation. Mr. Dugan made it clear that if any adverse effects, such as damage to oil-shale formations, contamination of either surface or ground water, unexpected seismic damage, or significant radioactive release from the detonations should result from preliminary phases, the entire project would be dropped by his company.[8]

As envisioned by the sponsors, the Phase II detonations, which were originally scheduled for 1974, were to be completely evaluated by 1975 or 1976. Upon the completion of well testing in these preliminary phases, it was anticipated that operating wells would be connected to a pipeline and Rio Blanco gas would become commercially available. With this portion of the work completed, government agencies and industrial sponsors could review and evaluate the results and decide whether to proceed with Phase III. Assuming a favorable analysis, Phase III could be started in the same year that Phase II was completed.

The purpose of Phase III, the last experimental step, would be to stimulate approximately 20 to 60 additional wells, or enough to justify the construction of a major natural gas pipeline. The sponsors viewed this segment of the plan as "the last or pilot plant step in demonstrating that natural gas can be commercially produced by nuclear stimulation."[9] Phase IV, predicated on the successful development of wells throughout the field in earlier phases, would systematically develop the entire field, and presumably result in a constant flow of natural gas from field to pipeline.

Even though Phase III of the project was expected to provide enough wells to supply natural gas for the new pipeline, the yield of natural gas from any well decreases with time as gas is drawn off and reserves become depleted. Then additional wells are needed to maintain a constant rate of flow. In Demonstration Program plans, CER cited a hypothetical case involving a pipeline at capacity for 25 years: "If, for instance, a constant pipeline capacity of 350 MMSCF [million standard cubic feet] is selected for the model, then a 58-well program is sufficient to provide the capacity to keep the line loaded for approximately three years. At the end of three years, additional wells would have to come onstream to keep the line at capacity."[10] Thus, full field development and maintenance of pipeline capacity would eventually necessitate one or more nuclear wells per section in the Rio Blanco unit. This means that a minimum of 148 wells, stimulated by three to five nuclear explosives in each well, would be required for full field development.

Again, assuming success up to this point, it would be only the beginning of natural gas production by nuclear means, because the Rio Blanco tests were to be the pilot plan for the use of the technology on a widespread basis. The basic hypothesis of all nuclear gas stimulation projects has been that an estimated 300 trillion cubic feet of natural gas in tight formations which is not economically recoverable by conventional means can be profitable if retrieved by nuclear stimulation techniques. This 300 trillion cubic feet of gas is located in diverse areas of the Rocky Mountain West and therefore clearly suggests that the Rio Blanco operations would only be a prelude to more and more frequent stimulation over a large geographical area.

Figure 6-3 shows the areas in the Rocky Mountain West which contain natural gas resources that are considered suitable for nuclear stimulation technology. The magnitude of the possible development anticipated by the Lawrence Livermore Laboratory

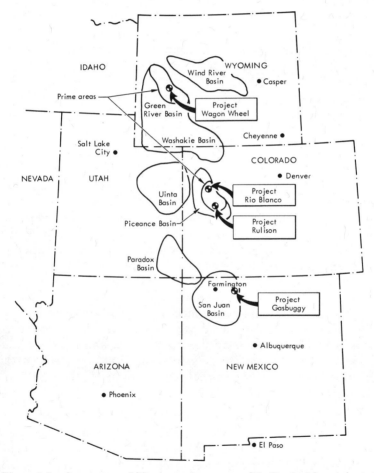

Figure 6-3. Low-permeability natural gas areas of the Rocky Mountain states. (B. Rubin, L. Schwartz and D. Montan, *An Analysis of Gas Stimulation Using Nuclear Explosives*, Lawrence Livermore Laboratory Report, May 15, 1972, p. 14.)

is illustrated in the *Addendum* to the *Rio Blanco Environmental Statement*, where it is estimated that 5,665 nuclearly stimulated wells would be required to fully stimulate the Piceance, Green River, and Uinta Basins in Colorado, Wyoming, and Utah. Some 2,400 of these would be in the Piceance Basin.[11] These figures do not even include the Paradox, San Juan, and Wind River Basins in other locations. An order of magnitude calculation suggests that a total of 10,000 to 15,000 nuclear stimulation shots would be necessary to fully develop the natural gas resources in the five-state area shown in Figure 6-3.

The delay in Phase I detonation from early 1972 to May of 1973 set the whole Rio Blanco time schedule back from that which was initially envisaged. Since then, a number of technical difficulties, disappointing preliminary test results, and political obstacles have emerged which cast doubt on the future of the Rio Blanco Project in particular and the Plowshare Program in general.

THE THEORY AND PRACTICE OF PHASE I

At the time of this writing in 1974, Rio Blanco Phase I preliminary testing and evaluation had been under way for a number of months. No complete and final analysis was available, but some preliminary results had been made public through the news media. The following discussion of Phase I is not an attempt to present an independent analysis of the preliminary results; it is a synopsis of the sponsors' estimates and predictions presented in the *Rio Blanco Environmental Statement* and *Addendum*, official documents, and public meetings prior to detonation. The only results that will be dealt with are those which have been released by the sponsors to the news media and in the official project publication, the *Rio Blanco News*. Tentative results of the Rio Blanco experiment and their implications for the future are considered in Chapter 7.

Unlike Projects Gasbuggy and Rulison, in which single nuclear devices were used, the Rio Blanco stimulation project used multiple nuclear explosives. Three 30-kiloton devices were set off simultaneously in the same well hole at depths of 5,840, 6,230, and 6,690 feet below the surface. Figure 6-4 depicts the configuration of cavities and fracture zones which were expected

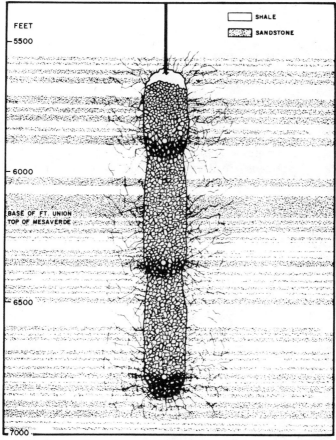

Figure 6-4. Expected configuration of the chimneys and fracture zone. (CER Geonuclear Corporation, *Project Rio Blanco Definition Plan, Vol. I, Project Description*, revised, Jan. 24, 1973, p. 23.)

to result from these simultaneous blasts. The detonations were expected to create three cavities which would interconnect to form one large chimney. The sponsors' analysis indicated that the vertical height of the Fort Union and Mesaverde geological layers would require the use of multiple explosives to stimulate the locked-in gas, because a single 90-kiloton explosive, although it could create a chimney with a larger total diameter, would produce a smaller vertical range than would three vertically spaced, 30-kilton devices. Because many of the underground gas-bearing sand layers do not naturally interconnect, the entire thickness of the formations must be stimulated to release most of the contained gas within a single well.

177

Table 6-1 lists the expected characteristics of the Rio Blanco chimney. The fact that the chimney and fracturing dimensions found in Projects Rulison and Gasbuggy were smaller than expected is reflected in the predictions for the Project Rio Blanco chimney listed in Table 6-1. The radial extent of fracturing predicted for Rio Blanco is compared to the 430 feet originally hoped for in the Project Gasbuggy chimney. But even the 250 feet may be optimistic, since an analysis of the Gasbuggy results indicates that the increased microfracturing may not extend to more than 2.5 times the cavity radius[12] (the predicted cavity radii are 72 to 75 feet in Rio Blanco). Hence, two and one-half cavity radii in the Rio Blanco well would be 180 to 188 feet. Predictions of gas production and economics were based on the assumption that the estimated chimney dimensions would be equalled or exceeded.

TABLE 6-1
PREDICTED CHARACTERISTICS
OF THE RIO BLANCO CHIMNEY

	Dimensions in feet		
	Upper	Middle	Lower
Detonation depth	5,840	6,230	6,690
Cavity radius	75	74	72
Chimney radius	83	81	79
Height of chimney above detonation point	250	240	240
Height of fracture above detonation point	310	300	300
Depth of fracture below detonation point	150	150	140
Radial extent of fractures from chimney center line	260	250	250

Source: CER Geonuclear Corporation, *Project Rio Blanco Definition Plan, Vol. I, Project Description*, Revised, Jan. 24, 1973, p. 22.

The principal reason for using several explosives in one well hole is to achieve stimulation over a large vertical interval in thick gas-bearing formations. In addition, though, it was expected that the interaction of the shock waves from the explosives would increase the extent of fracturing, especially between the devices, thereby enhancing the stimulation effect of the explosives.[13]

Predictions of reservoir performance in Project Rio Blanco were made by the AEC's geological consultants, H. K. Van Poollen and Associates. They were based on reservoir characteristics found in tests run on the Rio Blanco entry well (RE-E-01). Predictions for cumulative natural gas production from a nuclearly stimulated well in the Rio Blanco unit range from 10 to 37 billion cubic feet (bcf) over a period of 25 years. Van Poollen and Associates made eight separate predictions, varying their estimates of reservoir permeability, well spacing, and gas potential. One prediction of a total gas production of 20 bcf over 25 years, for example, is based on one well per 640 acres and the average permeability of the vertical interval.[14] The 20 bcf estimate given represents approximately 26 percent of the estimated gas-in-place in 640 acres in the Rio Blanco unit.

If we assume 20 bcf gas production from the test well in Project Rio Blanco, the natural gas yield per kiloton would be about five times greater than that of Project Rulison and more than seven times greater than that found in Project Gasbuggy. This dramatic increase would be due partly to more gas-in-place in the reservoir and partly to more effective stimulation. Table 6-2 compares the

TABLE 6-2
COMPARISON OF ESTIMATED GAS PRODUCTIVITY
FOR NUCLEAR STIMULATION PROJECTS

	Project Gasbuggy*	Project Rulison*	Project Rio Blanco	Project Wagon Wheel*
Explosive yield, kilotons	29	43	90	500
Estimated gas production, billion cubic feet	0.9	1.8	20	14-35
Billion cubic feet of gas produced/kiloton yield of nuclear explosive	0.031	0.042	0.22	0.028-0.07
Percentage produced of gas-in-place per section	4	6.6	26	7-17

*Predicted on the basis of 20 years.

predicted gas production per kiloton of nuclear explosive used and the percent of gas-in-place produced for each of the three completed projects and for the Wagon Wheel Project which is proposed for Wyoming.[15]

The predictions of reservoir productivity do not take into account gas quality or composition. The quality of natural gas as a heat source is based mainly on its methane content; the higher the percentage of methane, the better the quality of the gas. The Rio Blanco composition of the gas 90 days after detonation and immediately prior to production testing has been predicted to be about 26 percent methane, 34 percent carbon dioxide, 18 percent hydrogen and 22 percent steam.[16] As in the other tests, gas composition was expected to improve with production and approach that of the original field gas. Therefore, the value of the gas produced initially should be discounted because of its low heating value.

The contamination of Rio Blanco gas by radionuclides was expected to be much lower than that of either Gasbuggy or Rulison. As in Project Rulison, a three-month waiting period before re-entry followed detonation to allow short-life gaseous radionuclides and radioactive particles to decay. After the waiting period, the well was re-entered and production testing begun to determine the potential of the well. It was planned that up to 800 million cubic feet of gas would be withdrawn from the well during the initial testing period of 20 to 40 days. In previous experiments, water vapor produced with the gas was removed from the gas stream, the gas and water flows were measured, and the water vapor was reinjected into the gas stream before flaring. In Project Rio Blanco, the water was disposed of by injecting it into a nearby well to approximately the same depth as the nuclear well. This reduced the release of radioactivity into the local atmosphere during flaring. However, the amount of radioactive water produced was greater than anticipated and an amendment to the water-disposal permit had to be obtained from the Colorado Water Pollution Control Commission. The alternative of reinjecting the radioactive water vapor into the gas stream before flaring was rejected, even though CER officials had stated: "Meteorological and ecological studies indicate that even with the release of all gaseous radioactivity to the atmosphere, the maximum dose commitment to persons in the vicinity will be less than one percent of annual natural background radiation."[17]

Table 6-3 lists the predicted radioactive gaseous elements which were expected to be produced by the three Rio Blanco nuclear devices. Originally, the amount of radionuclides expected was classified information, and since the AEC has often based many of its calculations on "worst case" possibilities, the values presented here may be greater than those expected, and actual radionuclide production might well be somewhat less than the predicted figures.

TABLE 6-3
ESTIMATED RADIOACTIVITY PRODUCED AND RELEASED IN PROJECT RIO BLANCO DETONATION AND FLARING

Radionuclides	Total production (Ci)	Quantity assumed in flared gas (Ci)
Tritium	3,000	1,440*
Carbon-14	22.5	20
Argon-37	15,000	14,000
Argon-39	20	18
Krypton-85	2,000	1,800

*This value assumes that the water produced is flared with gas. If not, this quantity is approximately halved.

Source: Atomic Energy Commission, *Project Rio Blanco Environmental Statement,* WASH-1519, April 1972, p. 3-16.

After Rulison and Gasbuggy experience, the AEC developed an explosive specifically for gas stimulation. This type of explosive, referred to as a "Diamond" device, was successfully tested in the Miniata event in July 1971 at the Nevada Test Site.[18] Presumably, the explosives used in Project Rio Blanco were of this type. Important characteristics of the Diamond explosive were a small diameter (8 inches) and "minimum-residual tritium." Detailed information about the Diamond device was not released, because, as Mr. John S. Kelly of the AEC explained, "Due to security reasons . . . all materials information, including cost, are under the classified heading."[19]

181

Predictions of the seismic motion expected from the Rio Blanco detonations were subject to more uncertainty than those for earlier Plowshare tests because of the three-explosive configuration of the well and the complex interconnected chimney that was expected to be produced. The complex of seismic waves can be only partially modeled theoretically on a computer when a multiple explosive is detonated in varying rock layers. In essence, predictions of the seismic effects are based on the equivalent effects expected from one centrally located 90-kiloton device, with a maximum credible yield of 123 kilotons. Damage to nonindustrial surface structure, estimated on this basis, was predicted to be in the range of $51,000 ± $13,000. Buildings in the vicinity of the test site were structurally braced before the detonation in an effort to reduce structural claims.

The safety program undertaken by the sponsors to prevent any possible personal injuries required the evacuation of all people living within 7.2 miles of the project site. People living within 7.2 to 14.5 miles of the test site were asked to remain outside their homes or other buildings, and miners working within 53 miles of the site were evacuated from underground mines. People within 100 miles were warned to expect some ground motion. The total safety program carried out by the sponsors also included roadblocks to prevent injuries from rockfalls or landslides resulting from blast-created ground motion and rerouting of surface traffic in the vicinity of the test area.

Individuals who actively opposed Project Rio Blanco called attention at one time or another to several hazards associated with blast-engendered seismic activity. They argued that local inhabitants would not, and should not, be expected to agree to repeated detonations and the disruptions of daily life and threat of surface damage which they imply. The most compelling issue, in terms of commercial interests and potential alleviation of the energy crisis, was the contention that development of underground oil-shale deposits would be jeopardized by repeated nuclear detonations.

Though the oil-shale developers lost in their attempt to halt Phase I of the project, oil-shale operations are now going on in the Piceance Creek Basin on a trial basis without any apparent adverse effect from the nuclear detonations. However, further arguments of the type which preceded the Rio Blanco experiment can be expected, should there be further efforts to continue nuclear stimulations in oil-shale lease areas.

OIL SHALE OR NATURAL GAS: A CONTINUING CONTROVERSY?

The possibility of damage to oil-shale deposits in the Piceance Creek Basin by nuclear stimulation of natural gas reservoirs illustrates the problem if two potentially noncongenial technologies are introduced into the same area. The map in Figure 6-5 shows the known areas with potential for oil-shale mining, as well as the areas of gas-bearing formations where

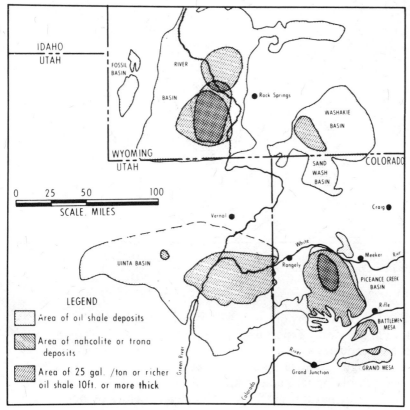

Figure 6-5. Oil shale areas in Colorado, Utah, and Wyoming. (U.S. Dept. of Interior, *Environmental Statement for the Prototype Oil Shale Leasing Program*, 1973, Vol. I, p. 1-3.)

183

nuclear stimulation is proposed. As mentioned previously, most of the gas-bearing rock formations have overlying oil-shale deposits. This is illustrated in Figure 6-6, which shows a geologic cross section of the Rio Blanco site. The conflict between oil-shale development and nuclear stimulation is focused basically on the timing of the recovery of the two resources. As oil-shale developers see it, if widespread nuclear explosions occur before oil-shale development, there might be irreparable damage to the structure of the oil-shale deposits.

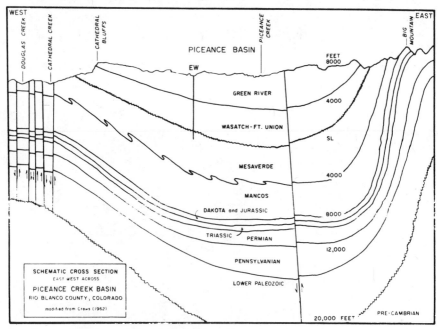

Figure 6-6. Schematic cross section of Piceance Creek Basin. (U.S. Atomic Energy Commission, *Environmental Statement, Rio Blanco Gas Stimulation Project*, WASH-1519, April 1972, p. A-9.)

The Oil Shale Corporation, better known as TOSCO, took the lead in providing expert testimony and studies to back up the oil-shale industry's basic opposition to nuclear stimulation technology. TOSCO spokesmen were careful to point out that they do not oppose nuclear stimulation *per se*, but do oppose the recovery of natural gas by nuclear means before or during oil-shale development operations. From their perspective, oil-shale development is of greater significance, both economically and practically, in adding to the nation's energy supplies than the recovery of natural gas in areas where both resources exist. Figure 6-7 depicts the depth of the oil-shale bed with the Rio

184

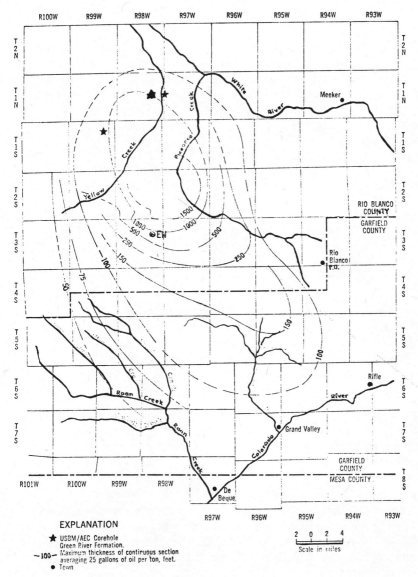

Figure 6-7. Isopach map of 25-gallon-per-ton oil shale, and the Rio Blanco gas stimulation area, Piceance Creek Basin. (U.S. Atomic Energy Commission, *Environmental Statement, Rio Blanco Gas Stimulation Project*, WASH-1519, April 1972, pp. A-29 and 2-21.)

185

Blanco gas field superimposed upon it. Since the Rio Blanco gas field lies in the middle of some of the richest known oil-shale deposits, it is clear why oil-shale proponents are concerned. Representatives of oil-shale interests therefore saw Phase I of Rio Blanco as a project which should be stopped before it began. If oil-shale deposits were developed before widespread nuclear stimulation of natural gas occurred, there would probably be little cause for conflict.

Nuclear blasting presents four major possibilities for damage to the oil-shale formations.[20] First, the explosion could open direct fractures spreading upward from the nuclear cavity to the shale zones. As Figure 6-8 illustrates, approximately 3,000 feet separate the uppermost nuclear explosive emplacement from the base of the oil-shale layer. Predictions of the extent of the fracture system based on Gasbuggy and Rulison data indicate that the maximum distance of outward fracturing in Rio Blanco will be 430 feet from the cavity wall. Thus, a rough minimum of 2,500 feet would lie between the expected outer edge of the nuclear fracture system and the base of the oil-shale deposits. This amount of separation would seem to rule out the possibility of direct fractures connecting the nuclear chimney with layers of shale deposits.

The second possibility is that blast shock waves might further disrupt already existing faults. A major disruption of a fault could weaken the entire geological formation, even to the extent of triggering a local earthquake. A detailed geophysical study of subsurface configurations in the Rio Blanco experimental area shows no major existing faults within about two miles of the entry well. Coring operations conducted by CER at the Rio Blanco experimental wells substantiated the conclusion that there are no major fracture systems within the entire geologic subsurface of the test area. It was concluded that the absence of significant faults precluded the weakening of oil-shale formations by blast-caused shocks.[21]

The third possibility is that shock waves would separate shale layers along either joints or bedding planes, or both simultaneously. To test the likelihood of this possibility, an analysis of the effect of shock waves on the bedding planes in the oil-shale deposits was carried out. The results indicated that "no fracturing could be calculated to occur even assuming the rock had no tensile strength at all and that the mismatch between the

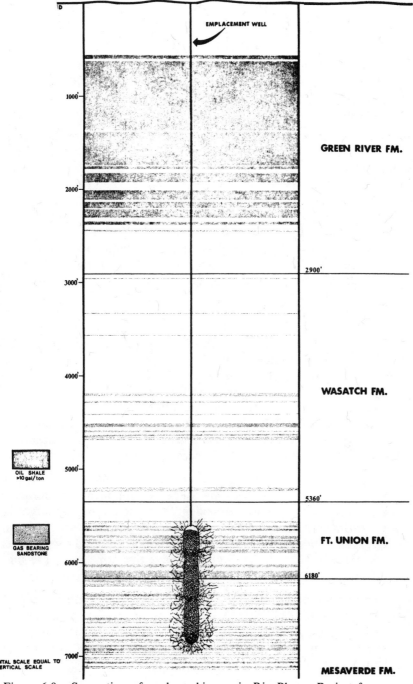

Figure 6-8. Separation of nuclear chimney in Rio Blanco Project from overlying oil-shale formations. (U.S. Atomic Energy Commission, *Environmental Statement, Rio Blanco Gas Stimulation Project*, WASH-1519, April 1972, p. 2-11.) Horizontal scale is equal to vertical scale.

properties of adjacent layers was two-fold greater than any measured mismatch properties in rock at the Rio Blanco site."[22] Since a body's tensile strength is its ability to withstand being pulled apart, an assumed tensile strength of zero means the only force holding the rock together and in place is the weight of the rock above it and the support of the rock below.

The fourth possibility for seismic damage could result from a spall effect. The term spall refers to the fact that when a force wave travels toward the ground surface, it is reflected by the surface much as light is reflected by a mirror. This force wave then travels downward and can separate layers of rock near the surface as it moves through them in its downward path. Calculations on the spall effect, assuming zero tensile strength in the rock, indicated that blast-related spall effect would not reach more than 350 to 400 feet below the surface. On this basis, the AEC concluded that the spalling effect would not reach oil-shale deposit zones.[23] In contrast, TOSCO geological consultants concluded that the spall phenomena could penetrate to 1,200 feet from the detonation point:[24] "Experience at the Colony Mine [the experimental oil-shale mine located a few miles from the Rio Blanco site] . . . indicates that there are regions of the oil shale that completely lack cohesion. Any tension exerted across these bedding planes will extend the regions lacking cohesion."[25] TOSCO consultants therefore concluded, contrary to the AEC prognosis, that Rio Blanco detonations could cause some separation of the shale layers above the nuclear emplacements and make oil-shale mining extremely difficult as a result.[26]

The AEC was, in short, not convinced by the evidence presented by oil-shale advocates. In spite of their confidence, the conflict in scientific opinion leaves some degree of uncertainty. Though the sponsors' predictions of no damage to oil-shale deposits was apparently borne out by the first Rio Blanco test, a large number of repeated detonations might not have this same benign seismic effect.

Ben Weickman of Superior Oil Company, in his testimony before the Governor's Oil Shale Advisory Committee, pointed out that there existed both technical and economic considerations if oil-shale and nuclear stimulation technologies were carried out simultaneously. According to Weickman:

Nuclear explosions, because of the unpredictability, cause fear and apprehension in most people. This fear and apprehension among the financiers of the oil shale industry could cause a complete breakdown of potential financing of the oil shale industry. Because an oil shale industry would require tremendous amounts of money, and because a good deal of this money would have to be committed prior to a determination as to the amount of damage that could be imparted on the oil shale industry by nuclear explosions in an adjacent area, it is most likely that oil shale financiers will elect not to invest in the oil shale industry until this uncertain factor is eliminated.

In an attempt to investigate this point, letters were sent to five banks with current commitments to provide financing for the oil industry.[27] Four replies were received from vice-presidents of the concerned banking institutions. All of these bank officials generally felt that there were more serious obstacles to financing oil-shale ventures than fear and apprehension about the use of nuclear explosives. A. C. Gueymard, Senior Vice-President of the First National City Bank of Houston, Texas, for instance, wrote that "the fact that nuclear explosions cause fear and apprehension is not quite as much of a deterrent as that of the ecologists and other hairbrained individuals who overemphasize situations and who are, also, but poorly acquainted with the real facts. A lender would be apprehensive, certainly, about financing an oil-shale venture where the situation existed to the effect that environmentalists cause a delay or complete shutdown of the operations." Other responses mentioned such obstacles as unproven technology, unresolved environmental problems, and the fact that the economics of oil-shale recovery are still questionable. None of the four respondents felt that any or all of these obstacles implied "a complete breakdown of potential financing."

Officials of TOSCO and representatives from other oil-shale development companies were primarily concerned about the effects of full field development on oil-shale deposits rather than about the Phase I explosions of Project Rio Blanco. They essentially admitted that Phase I might not cause any damage to overlying shale formations, but the cumulative effect of many blasts could well produce enough damage to prevent oil-shale extraction. As Mr. Weickman put it in completing his testimony

189

to the Governor's Commission, "Responsible calculations indicate that the same amount of gas can be recovered over a twenty-year period by more intensive well drilling and conventional techniques at a cost of about half that of a nuclear stimulation project."[28] Logic, then, to Mr. Weickman, dictated that natural gas in the Piceance Creek Basin should be recovered by conventional means, since such a technique would be profitable and at the same time eliminate potential nuclear risks to the oil-shale industry.

The sponsors of Project Rio Blanco did not attempt to refute claims that oil-shale development offers far greater economic potential than nuclear stimulation of natural gas. Rather, they consistently maintained that natural gas in the Rio Blanco field could be stimulated by nuclear means before, or simultaneously with, the development of oil shale without damage to the latter. Mr. Dugan, of the Equity Oil Company, noted several times that if any damage were inflicted on oil-shale deposits by Phase I, his company would drop out of the entire project. Since Equity holds significant leases on oil-shale lands in the Piceance Creek Basin, it clearly has twin economic stakes in the area. Quite logically, profitable economics will determine the ultimate commitment of company resources, and perhaps in the long run the future of all nuclear stimulation projects.

AGAIN, THE COURT SCENE

As the appointed time for the detonation of the Rio Blanco devices approached, history seemed to repeat itself. As in the days preceding the Rulison experiment, many of the same technical, environmental, and public health uncertainties became apparent, and the same kinds of last-ditch efforts to halt the detonation occurred in the courtroom. In the Rio Blanco case, however, a suit to halt the test was sought through the legal jurisdiction of the State of Colorado in the Denver District court of Judge Henry Santo, rather than through federal jurisdiction as in the case of Project Rulison. In both cases, however, the plaintiffs were environmental advocates who were ultimately denied the injunctive relief they sought, but who eventually found some grounds for optimism in the final verdict.

190

The Rio Blanco suit was initially filed on May 1, 1973, somewhat more than two weeks before the scheduled detonation date. The plaintiffs included Citizens for Colorado's Future (the group which had successfully led the petition and referendum drive to move the 1976 Winter Olympics away from Colorado), the Colorado Open Space Council, and the Environmental Task Force. Individual plaintiffs included Jerry Cook, a landowner from Rifle; Paul N. Brown, a Grand Junction realtor; and three physicians, Dr. Harold Harvey of Aspen, Dr. Charles Gaylord of Denver, and Dr. Edward Cheney, also of Denver. David Engdahl, then a faculty member of the University of Colorado School of Law, was chief counsel for the plaintiffs.[29]

Professor Engdahl, in filing the suit for the plaintiffs, basically argued that the Colorado Water Pollution Control Commission had failed to prove beyond a reasonable doubt that the Rio Blanco detonation would not pollute state waters, and it had not been shown that the experiment would be in the public interest. At the time Engdahl filed the suit, he also sent letters to Water Commission members, requesting a postponement of the effective date of the Rio Blanco permit. Among other things, Engdahl noted that the Commission had taken the desirability of enlarging the commercial natural gas supply as justification for the project without showing that this *one* explosion could significantly enlarge that supply; he also pointed out that the Commission itself had substantial doubts that the technology could safely be applied on a sufficiently large scale to enhance that supply significantly. Engdahl further noted that under federal law it was not even permissible to commercially use natural gas stimulated by nuclear means and that efforts over the past several years to legally permit such use had failed.[30] Thus, he contended, the Commission's rationale for allowing the blast on the basis of enlarging usable natural gas supplies was questionable on a number of grounds.

In spite of the fact that the Colorado Water Pollution Control Commission had voted on April 17 to issue a permit allowing the detonation, no formal permit had been certified by May 1. For this reason, Judge Santo ruled initially that the case was not yet "properly before the court," because before the Commission's action in granting a permit was specifically certified to the court, there was no administrative action to review.[31] But after the Colorado Water Pollution Control Commission's proceedings were formally delivered and presented to the court, the case could be heard. During this initial brief hearing, William Schoeberlein, an attorney for CER, said that he would file motions in answer to the plaintiff's allegations the following week, but added that from his client's standpoint, the entire legal proceeding would become moot after May 9, the date that the Project Rio Blanco nuclear devices would be fixed in place underground. Judge Santo was not impressed with this argument, stating that once the case was properly before the court, the timetable for the project would not be a factor which would influence his decision.

When the record of the Commission's actions was formally presented to the court, there were some preliminary legal questions to be resolved before the merits of the case could be considered. Again, as in the Rulison litigation, the plaintiffs felt that they had won some skirmishes, but lost in the end. The question of the plaintiffs' standing to bring the suit was brought up by the Assistant Attorney General of Colorado, William Tucker, who was representing the Water Pollution Control Commission. He argued, although attorneys for CER did not join him in this contention, that the suit should be dismissed on the grounds that the plaintiffs lacked requisite legal standing. Judge Santo, in a broad interpretation, disposed of this motion by asking: "Are the plaintiffs residents of the state who would be affected by potential water pollution?" Given an affirmative answer, the Judge denied the motion for dismissal. This, in effect, established a precedent which grants citizens of Colorado the right to sue in state court to protect their interest in pure and usable water.

A second and thornier preliminary question concerned the defendants' argument that the Denver District Court lacked jurisdiction over the case because of the federal sponsorship of Project Rio Blanco. This argument rested primarily on the legal principles of federal supremacy and preemption, or on the doctrine that federal law is supreme and preempts a state's legal authority in specified areas. Mr. Engdahl, as the plaintiffs' attorney, argued against this contention by presenting a complex constitutional argument based on the concept that the doctrines of federal supremacy and preemption have been misinterpreted since pre-World War II years and that they apply only to certain types of federal projects carried on under specific conditions.

192

Essentially, he claimed that federal agencies and allied private contractors are protected by these doctrines only when they are engaged in activities covered by constitutionally delegated federal powers and when the private contractor is rendering a specific service to the government. The Rio Blanco case, Mr. Engdahl argued, did not come under constitutionally enumerated powers, and CER was not rendering a service, but rather was receiving one from the federal government in the form of AEC technical expertise and in the receipt of the nuclear devices.[32]

Judge Santo resolved this issue, but not on constitutional grounds nor on the basis of whether Congress did or did not intend to preempt state control. Instead, he relied primarily on the fact that Colorado is an "agreement state" in respect to the AEC, and to a lesser extent on the contract between CER and the AEC. The Judge cited Article 26 of the AEC-CER contract as speaking for itself in unequivocally stating that "CER Geonuclear Corporation shall procure all applicable and necessary permits or licenses and abide by all applicable laws, regulations and ordinances of the United States and of the state, territory and political subdivisions in which the work under this contract is performed."[33] Of even greater importance to Judge Santo's dismissal of the defendants' claim, however, was the fact that Colorado became an "agreement state" in respect to the AEC on January 16, 1968. Through this action, the regulatory authority of the AEC in Colorado was discontinued as far as "byproduct materials, source materials and special nuclear materials in quantities not sufficient to form a critical mass" were concerned. As Judge Santo saw it, CER specifically recognized the authority of the State of Colorado by the act of applying for a permit from the Water Pollution Control Commission. The question of the production of nuclear byproduct materials, in the Judge's view, was inextricably interwoven with the question of public health and safety, which is clearly a matter of state concern.

In preliminary objections, CER raised an additional jurisdictional question in arguing that the Colorado statutes which created the Water Pollution Control Commission and defined its authority were unconstitutional. This argument was based on the contention that the provision which requires the Commission to find "beyond a reasonable doubt" that a proposed activity is justified by public need is "an unlawful delegation of legislative authority," and is therefore unconstitutional.[34] The counsel for the defendants argued that this legislation provided no standards or guidance by which to interpret "public need," and that only the State Legislature, not a commission created by it, can constitutionally determine such a need. Did "public need" mean solely the needs of Colorado citizens, or the public need for an increase in natural gas reserves, or an even broader need for new sources of energy?[35]

Judge Santo dealt with this issue by referring back to his earlier ruling on the question of federal supremacy and preemption and the decision that a Colorado statute requiring an administrative agency to make a determination of "public need" was not in conflict with constitutional law or congressional intent. The Judge then cited applicable Colorado statutes defining "waters of the state," pollution and wastes, and the powers and duties of the Water Pollution Control Commission as proof of adequate standards and guidelines.[36] Thus the interpretation of "public need" was placed within the relatively narrow context of water purity. After consideration of all relevant statutes and amendments, the Judge concluded that the term " 'public need' must be applied in the light of attendant circumstances and is not to be used in a limited sense but for the interest and safety of the State in the light of the statutory purpose which is clearly enunciated by the statute itself." "Public need" thus was interpreted as a relative term which must be applied on an *ad hoc* basis. The broad legislative grant of authority to the Commission to determine "public need" without further definitions, standards, or guides is perfectly permissible "to fulfill the aim of the statute."[37]

Once these preliminary arguments and jurisdictional issues were resolved, the court was in a position to rule on the merits of the case. Under the Colorado Administrative Code, a court can review agency actions to assure that the agency has acted in accordance with due process and has not exceeded either constitutional or statutory authority. Thus, in such cases, the court does not consider new evidence, but only matters that had been considered by the agency; in this case, it was those considered by the Water Pollution Control Commission before the issuance of the permit for the Rio Blanco detonation.[38]

In their attempt to obtain a favorable decision, the burden of proof rested with the plaintiffs. In essence, they had to prove that they were adversely affected by the action of the Commission in allowing detonation of the Rio Blanco device. In addition to a permanent injunction to prevent Phase I of Rio Blanco, the plaintiffs asked for a similar injunction against any other nuclear detonation in the State without prior lawful action of the Commission.[39]

The plaintiffs made their case on the basis of four major contentions. In the first of these, it was alleged that the Commission had not made its own investigation, as required by law, before scheduling a public hearing on the issuance of the permit. Implicit in this argument was the point that an investigation and a hearing are not one and the same kind of action. Judge Santo reviewed Commission staff memos and the testimony from state and federal officials and independent witnesses that had been presented at the hearings; he was persuaded that an investigation had, in fact, been held.

The second point raised by the plaintiffs was based on the fact that Commission findings were not unanimous. Therefore, they claimed, the burden of proof was placed on the Commission to show that the detonation was clearly based on public need and that there would be no resulting pollution or significant migration of radioactive water. Judge Santo found that there was a legal quorum of members at the Commission meeting on April 17, 1973, and that nine of the eleven Commission members had voted to approve the permit. The issue was resolved on this point and again went against the plaintiffs.

A third argument advanced was that the Commission's formal findings and decision of May 4, 1973 *nunc pro tunc*, or, as entered into on April 17, 1973, were defective and of no validity. The reasoning behind this point was that the Commission proceedings which were certified to the court on May 4 included findings that were not available before the lawsuit was filed on April 25. Mr. Engdahl pointed out that on May 2 he had asked whether there were any findings by the Commission and was told that there were none available at that time. Yet on May 4, the Executive Committee of the Commission had issued its formal permit to CER; this was the same day on which the formal Commission findings were released. The plaintiffs contended that these findings consisted essentially of a verbatim copy of the Colorado State Board of Health report on the project and a hastily appended clause at the end stating that there was no reasonable doubt on the possibility of any significant movement of polluted waters. Further, it was alleged that the last-minute Executive Committee action was based, at least in part, on three telegrams and a notarized letter. In the plaintiffs' view, this procedure was an unlawful way of conducting Commission business. Judge Santo again ruled against the plaintiffs, indicating that *nunc pro tunc* orders are common and acceptable and that the telegrams and letter were effective and valid as personal signatures. It was, in other words, not required that the Commission be formally sitting as a group to take the final decision in authorizing the permit.

The concluding point in the plaintiffs' argument was the allegation that the Commission had not proved beyond a reasonable doubt that there would be no water pollution caused by the blast. Nor, they contended, was there adequate justification of the need for carrying out the project. The court's conclusion was that the Commission was well aware that the burden of proof rested with CER to demonstrate that there was no reasonable doubt as to public need and safety requirements before detonation could occur. Judge Santo then considered whether there was substantial evidence that the Commission's findings were adequately supported on the basis of the record. His review of the record led him to conclude that there was substantial evidence presented by competent persons which supported the issuance of the permit on grounds of both safety and public need. Although he agreed with the plaintiffs that the flaring of the natural gas produced by nuclear explosives could not be separated from the detonation itself, the issue before the Commission, and now before the court, was one of potential water pollution. The issue was not that of the release of potentially dangerous radioactive material through flaring. Having lost on all counts, the plaintiffs then sought an injunction to prevent detonation pending submission of an appeal. In this endeavor, they were equally unsuccessful.

Thus, once again, persons seeking environmental protection as they saw it, had their day in court, but failed to achieve the desired result. Nonetheless they felt, as did their counterparts in the Rulison cases, that they had made progress. Judge Santo, in their view, had set an important precedent in holding that Colorado citizens could bring suit in Colorado court to protect the State's water from possible pollution stemming from a federal project. The fact that the State could and did assume jurisdiction was gratifying to those who were pressing for state control of projects clearly affecting the State and its citizens. And since Judge Santo had specifically noted that he considered detonation of the explosives and the subsequent flaring of the gas to be integrally related, the legal basis was established for judicial review of any additional permits for flaring which might be requested.[40] As it happened, the sponsors did not flare all radioactive water with the gas, as it was agreed that most water would be disposed of underground. This fact, however, does not detract from the Judge's ruling, which made it clear that future project operations would be subject to the judicial scrutiny of the State of Colorado.

A few hours after Judge Santo's decision was given, a spokesman for Citizens for Colorado's Future announced that although the plaintiffs felt they had adequate grounds for appeal, they would not do so because appeal would be too costly. The initial suit was within the realm of possibility, given relatively low costs and the donated time of counsel. An appeal, however, would involve a number of additional costs which could not be met by the plaintiffs.[41]

As suggested at one time by Senator Floyd Haskell, a suit could have been brought in Federal Court on the grounds that there were procedural omissions by the AEC under the terms of the National Environmental Policy Act of 1969. This choice, however, would have involved exceedingly heavy costs which were beyond the financial capabilities of locally supported groups and individuals. It was, in any case, perhaps a more beneficial move from the point of view of the environmental advocates, to challenge Rio Blanco in State Court. In affirming the right of citizens to sue in State Court to protect their interests, the central issue of state jurisdiction *vis-à-vis* the federal government was squarely met. Officials of CER considered the possibility of appealing Judge Santo's ruling in an effort to overturn his decision on state jurisdiction; this step was not taken, perhaps because the adversary would have been the State of Colorado. In any case, it seems clear that both the State and a number of its citizens are equally anxious to reaffirm the right to exert control over conditions of life within the State.

MAY 17, 1973: PRELUDE OR FINALE?

With the last legal roadblock to detonation cleared by Judge Santo's decision, and all three bombs cemented in place, the sponsors made plans to set up their own roadblocks to seal off rail and auto traffic in preparation for detonation. The detonation was at one time scheduled for May 19, a Saturday, to avoid blasting at a time when children would be in the nearby Rock School. However, detonation day was moved up to May 17, a Thursday. The Federal Aviation Authority had stipulated that detonation should occur during daylight hours, prior to 9 a.m. or after 4 p.m., so that only minimal rescheduling of routes of peak-time air flights over the area would be required. In spite of these stated preconditions, the detonation occurred at 10 a.m. on Thursday, May 17, 1973.[42]

By the detonation time, fifty residents within 7.5 miles of the test site had been evacuated to assembly centers. All people living within 7.5 to 14.5 miles of the site were requested to remain outside of and away from their homes or other buildings. Highway road crews were standing by to clear away any fallen rocks, and air space for a six-mile radius around the site was closed to private aircraft. Monitoring and environmental surveillance equipment was in place and ready for operation. Colorado Governor Love, Dr. Ray of the AEC, and several members of the Joint Committee on Atomic Energy, including Senator Dominick of Colorado, were in a helicopter above the detonation site to observe the detonation from the air. Four hundred other people were on hand at the designated observation point. At the entrance to the observation post, individuals from Environmental Action of Colorado, Delta County Citizens Concerned about Radiation, and a Boulder environmental group protested by way of a flaming sign which asked, "Who the hell gave permission for these blasts?"[43]

Clearly, permission was given, and given by an assortment of federal and state agencies. The three nuclear devices were detonated simultaneously at depths of 5,840, 6,230 and 6,690 feet, producing an explosion with a force four and one-half times larger than the Hiroshima atomic bomb and over half again as powerful as the Rulison blast. Specifically, the Rulison shot was equivalent to 43,000 tons of TNT in comparison to the Rio

Blanco equivalent of 90,000 tons. Seismic shock waves from the blast registered 5.5 magnitude of shock on the Richter scale, as recorded at the Colorado School of Mines at Golden, on the eastern slope of the Continental Divide. The fact that the shock was "no greater than the most severe reported during earthquakes originating from the Commerce City-Darby Geological Fault" on the eastern slope of the Divide may not have been particularly reassuring to Colorado residents.[44]

The sponsors announced shortly after the test that a preliminary analysis indicated that all three devices had detonated successfully, that there was no radiation released at the surface, and that seismic activity was well within predicted levels.[45] More specific analysis of test results was of course not possible until the well was re-entered and the gas flared in preliminary production testing. As was the case in the Rulison experiment, re-entry was scheduled for a number of months after detonation in order to allow time for the short-lived radioactive by-products of the blast to decay.

Before the scheduled September re-entry was undertaken, however, CER held a seminar on the results of the project to date. In the seminar, called "D + 30 Day Report," held in Grand Junction in July 1973, the presentation focused on summarized data and analysis completed through June 16, 1973. More than 70 people—scientists, city, state and federal officials, industrial representatives, and area residents—were told that "the detonation phase . . . has been a success and that the effects of detonation were close to predictions."[46]

It was made clear to the seminar group that the most critical part of the test—namely, determination of the well's productivity and the quality of the gas—was yet to come. However, as of June 16, a number of tentative conclusions could be drawn. Radiation measurements continued to show no radiation increase above normal background levels. The spall effect was described "in the form of a shallow dish with a depth of about 350 feet and a radius of less than 24,000 feet." It was concluded on this basis that there was no damage to oil-shale deposits which lie about 400 to 2,300 feet below the surface in the test area. Ground-motion measurements averaged from 50 to 60 percent of the predicted levels, and the aftershocks which did occur were mainly concentrated within a two-hour period following detonation and were "hundreds of times smaller in equivalent yield than the detonation." In addition, the underground fault structure remained stable with no indication of movement of adjoining fault blocks. There were some rockfalls following detonation which blocked roads temporarily and caused some damage to a mine 18 miles from the blast site.

Although water-sample analysis was incomplete by June 16, there was no significant difference in chemical composition of samples taken before and after detonation. Some springs as far as 10 miles from the site which had been dry for a number of years began to flow, probably in large measure due to blast-engendered shock waves. Surveys of the blast's effect on vegetation and wildlife showed no evidence of injury to deer or cattle, but some loss and alteration of habitat of small animal and bird species were caused by rockfalls. Also some loss of trees, shrubs, and grasses was caused by rock and earth slides.

Surface structural damage reported by mid-June was less than predicted, seemingly because of less than expected ground motion and the extensive structural bracing program that was done prior to detonation. A year and a half later, in January 1974, 167 claims had been filed for a total amount of $39,245.90. The largest claim amounted to $4,181.50 and the smallest was $5.20. The largest single residential claim was $2,229.91.[47] In contrast to Rulison claims settlement, in which the claims office was closed a year after detonation, the Rio Blanco claims office continues to remain open in Grand Junction.

In mid-June of 1973, the technical promise of the Rio Blanco test thus seemed reassuring. The political climate following detonation, however, was somewhat less than reassuring, as indicated in a statement by Dr. Ray of the AEC on May 18: "The decision about what will be done with this technology will not be up to the AEC. Any decision . . . will have to be made in Congress with the involvement of the public." Governor Love, at the same time, noted that he "couldn't presently imagine the circumstances in society where . . . [nuclear stimulation] would be acceptable."[48] John Vanderhoof, who shortly became Governor following John Love's departure to become President Nixon's Energy Chief, said that he probably would not have allowed detonation on the basis that oil shale has a higher energy potential than gas. Although the power of the Governor to veto an AEC project does not seem to be legally or politically clear, Hal Aronson, the Vice-President of CER, seemed to have no doubt that Governor Love could have vetoed the project, had he felt that such action would be in the public interest.[49]

In the spring and summer of 1973, the public interest was otherwise variously interpreted. The City Council of Aspen, Colorado, went on record as opposing the proposed commercial use of "radioactive gas" supplied by Colorado Interstate Gas Company from Project Rulison.[50] The *Denver Post* and the *Rocky Mountain News*, the two largest circulation dailies in the state, editorially opposed any further Rio Blancos.[51] The *Post* environmental editor, Dick Prouty, especially concerned that the public should be consulted as to the public interest, suggested that the whole issue should go on the ballot in a public referendum.[52] Representative Pat Schroeder of Denver not only supported his idea, but also took a somewhat lonely anti-Rio Blanco stand in Congress. As a protest over the refusal of the House of Representatives to accept an amendment to divert $38 million from the Plowshare Program, she voted against the entire $2.4 billion AEC funding measure. Other members of the Colorado delegation in the House, however, voted in favor of the authorization.[53]

Anti-Rio Blanco sentiment notwithstanding, the long hot summer of 1973 was sure to be followed by a not-so-hot winter, and there were people both in and out of the State who insisted that the energy crisis must be placed into realistic perspective. Nuclear technology, then as now, was seen by many as the solution to the nation's energy needs. One of the best-known advocates of nuclear stimulation was then, and still is, Dr. Edward Teller, who played a significant role in the decision to develop the hydrogen bomb. Speaking to a Boulder audience in June 1973, Dr. Teller said that a large number of nuclear explosions would be necessary to stimulate natural gas, but that "fuel from Colorado could solve the nation's energy crisis."[54] Although the dimensions of the energy crisis have become increasingly apparent to all citizens, Dr. Teller's reference to the necessity of up to a thousand nuclear detonations in the State was not persuasive to all citizens. If it was true, as Dr. Teller suggested, that the people of the State would decide whether nuclear blasts were to continue, then it was up these same people to take the initiative and make their sentiments known. This is

201

precisely what the voters of Colorado did in passing a referendum measure in the fall of 1974 to prevent further underground nuclear explosions in the State without a prior affirmative vote of the people. Thus, it appears now possible that the voters of the State will be asked to make a value judgement on the need to produce energy as opposed to the need for present and future protection against radioactive contamination. The people, in short, may become the arbiters of the underground nuclear explosion decision-making process, a process that has heretofore been beyond their reach.

RIO BLANCO: AN EPILOGUE

In the period between the detonation of the three Rio Blanco nuclear devices and the election in November of 1974, the people of Colorado became aware that the underground nuclear stimulation experiments on the Western slope had produced some unexpected results that apparently neither the AEC nor its industrial sponsors had foreseen. The drilling of the reentry hole, designed to determine quantitatively how successful the nuclear stimulation had been, was delayed for two months by a blockage of the reentry route which had been produced by the nuclear blasts. However, the Rio Blanco well was successfully reentered in November of 1973 and production tests were carried out.

After ten days of production testing, which included the flaring of approximately 30 million cubic feet of gas per day, the well was shut down in order to allow the gas pressure to build up again. The same procedure was followed again in late January, 1974. This production testing indicated that of the tracer gases placed in each nuclear device, only gas from the top device emerged. This led to the conclusion that all gas produced so far had come from the fracture zone nearest the surface. In short, it appeared that the three nuclear devices had not succeeded in connecting the top, middle, and bottom fracture zones.[55]

Following the January production testing it became apparent that the well used to dispose of the tritiated water produced with the gas had developed a small leak in the pipe used to force the water into the disposal well. A CER official, when questioned about the leak, said that he assumed that the radioactive water would flow down into the well but that he had not yet looked closely at the injection system. An AEC representative from the Lawrence Livermore Laboratory acknowledged that there was a leak but said that the water "... contained only slightly more than background levels of tritium."[56] However, when the Environmental Protection Agency analyzed the gas samples, it was found that the gas contained 1000 times the normal background radiation levels.[57]

By mid-June of 1974 the decision was made to drill a second hole in an effort to penetrate the cavity created by the middle explosive in the three-tiered Rio Blanco design. In April, 1974, according to Senator Floyd Haskell, the AEC indicated that it would need another $375,000 to complete evaluation of the project, but by mid-June the figure had risen to $1.4-million.[58] On June 20 the projected cost of this second effort was placed at $2-million in the *Denver Post*.

On the basis of the February, 1974, production test, the project sponsors concluded that there were unexpected increases of radioactive tritium in the water produced by the Rio Blanco detonations. The additional amounts of tritiated water were apparently produced because the temperatures below ground followng the blast were much higher than expected. As a result, the sponsors had to request a permit from the Colorado Water Quality Control Commission to increase both the volume of radioactive water and the amount of radiation to be disposed of underground.

In the interval between the winter of 1973, when the sponsors sought a permit to detonate the Rio Blanco devices, and June, 1974, the State Legislature passed a bill changing the name of the Water Pollution Control Commission to the Water Quality Control Commission. In addition, this legislation gave the State Department of Public Health statutory authority to regulate water quality control standards in the state. The original permit for underground disposal of tritiated water allowed for disposal of 24,000 barrels of water with a radiation content of 50 to 150 curies. The sponsors now asked the Commission for a permit allowing for disposal of 34,000 barrels with an estimated radioactive content of 459 curies.[59]

In an attempt to resolve the dilemma of radioactive waste disposal, a hearing was held at which various experts were asked to testify. A State Hearing Officer from the Administrative Division of the Department of Health concluded after the completion of testimony by independent experts that the alternatives to allowing subterranean disposal seemed undesirable: One would involve trucking the contaminated water to Nevada for disposal, the other, evaporating the water through flaring into the atmosphere. During the hearing, Dr. Maurice Major of the Colorado School of Mines said that the possibility of hazard from earthquakes resulting from underground burial was ". . . so slight it need not be considered," and Dr. Henry Van Poolen testified, "The risks involved in the proposed injection of tritium-bearing water, if any, are minimal" and "No better method of disposing of the tritium radioactive hydrogen has been advanced."[60]

The Water Quality Control Commission adopted the findings of the hearing examiner in mid-October, 1974. The Commission members were satisfied that the Continental Oil Company, the parent company of CER Geonuclear Corporation which had taken over Rio Blanco operations, had proved beyond reasonable doubt that potable water would not be polluted by subterreanean disposal. The Hearing Examiner for the case, however, noted in his findings that the Project sponsors, including CER and the AEC's Lawrence Livermore Laboratory, were in error in their predictions of the amount of radiation, the volume of water, the blast-incurred temperatures and the direct results of the simultaneous blasts themselves. Nevertheless, upon the affirmative vote of the Water Quality Control Commission, the Department of Public Health issued the requested revised permit.[61]

With these difficulties resolved, the project sponsors proceeded with production testing on the second reentry well for which drilling had been begun in June. Unexpectedly, on November 6, it was announced that another potentially harmful radioactive element, cesium-137, was contained in the natural gas and water emanating from the well. This contamination was a complete surprise for the public because the Environmental Impact Statement on Rio Blanco had stated specifically that cesium-137 and strontium-90 would be trapped in solidified molten rock at the bottom of the nuclear chimney and could not escape. A later amended version of the Impact Statement indicated that some of the material would be deposited on underground rock surfaces, but that neither of the substances would be able to escape from the nuclear-created cavity. When questioned about this radioactive contamination, Ben McCarty of the AEC Plowshare Office in Washington said that the AEC did not know why the cesium was produced with the gas.[62]

Hard on the heels of the discovery of cesium in the gas came an announcement from the State Department of Public Health that strontium-90, another radioactive material linked as a causative factor in cancer and leukemia, had also escaped from the nuclear chimney. Moreover, the second production test provided further confirmation that the three underground cavities had failed to interconnect. Since there was no underground connection and three separate and connected cavities had been formed, the original well was virtually useless for drawing off accumulated natural gas. Thus, the multiple-blast technology had, in effect, failed to live up to its expectations.

Once again, the Department of Public Health and the Water Quality Control Commission were faced with the problem of disposing of radioactive materials. Charles Gouge of the Commission commented that he hoped that this was the ". . . last surprise that the Commission . . . was going to be hit with." Director of Natural Resources Tom Ten Eyck concurred with the observation that "the people's previous reliance on the infallibility of our scientists has certainly been shattered [by the disclosures] and was probably the cause of the passage of Amendment 10."[63]

The Rio Blanco story is thus now largely complete. The nuclear well, like its Gasbuggy and Rulison predecessors, has been sealed and there appears to be no immediate plan for its reactivation. Testing, however, is proceeding on a new Rio Blanco well stimulated by massive hydraulic fracturing. Though results at this time are not conclusive, the conventional technology appears to be promising. The question remains as to why the more conventional alternative to the nuclear stimulation of deep underground natural gas was delayed so long. The answer, it would seem, lies in the technical, political, economic, and legal ramifications of the Plowshare Program itself.

NOTES

[1]Equity Oil Company, *Special Report to the Stockholders*, Nov. 18, 1972, p. 3.

[2]Since federal legislation currently prohibits the sale or independent commercial use and development of nuclear explosives, the AEC would be responsible for supplying the nuclear devices and related safety services. This figure is thus exclusive of any costs of the explosives and preliminary research and development done by the AEC.

[3]Equity Oil Company, *Report to Stockholders*, p. 4.

[4]CER Geonuclear Corporation, *Project Rio Blanco Definition Plan, I, Project Description*, Feb. 14, 1972, p. 4 (available from U.S. Bureau of Mines Library, Denver, Colorado).

[5]*Ibid.* See also Equity Oil Company, *Report to Stockholders*, p. 6.

[6]CER, *Project Definition Plan, I*, p. 3.

[7]CER Geonuclear Corporation, *Demonstration Program, I*.

[8]Conversation on Oct. 10, 1972, between Paul Dugan and David Hunter of the Technology, Law and Politics Research Project staff at the University of Colorado.

[9]CER, *Demonstration Program*, p. 10ff.

[10]*Ibid.*, p. 13ff.

[11]U.S. Atomic Energy Commission, *Environmental Statement, Rio Blanco Gas Stimulation Project*, Addendum, Washington, D.C. (WASH-1519), March 1973, p. E-72.

[12]I. Y. Borg, "Micro Fracturing in Postshot Gasbuggy Core G B-3," *Nuclear Technology, 11*, July 1971, p. 379.

[13]M. D. Nordyke, "The Use of Multiple Nuclear Explosives for Gas Stimulation," *Nuclear Technology, 11*, July 1971, p. 303.

[14]H. K. Van Poollen and Associates, *Projected Reservoir Performance, Rio Blanco Project*, NVO-38-33, March 1972.

[15]*Environmental Impact Statements, Project Rio Blanco and Project Wagon Wheel*, (WASH-1519, WASH-1524).

[16]AEC, *Environmental Statement, Rio Blanco*, pp. 3-15.

[17]CER, *Project Definition Plan, I*, p. 33.

[18]*Annual Report to Congress of the Atomic Energy Commission for 1971*, Jan. 1972, p. 140.

[19]"Successful Miniata Shot Raises Hopes for Faster Development of Gas Stimulation Explosives," *Nuclear Industry, 18*, No. 7, July 1971, pp. 27-32.

[20]This discussion of the potential damage to oil-shale deposits is based mainly on testimony presented by The Oil Shale Corporation and by Superior Oil Company before the Governor's Special Advisory Committee on the Rio Blanco Nuclear Stimulation Project, October 1971, and before the Informal Public Hearings on Project Rio Blanco conducted by the U.S. Atomic Energy Commission at Meeker and Denver, Colorado, March 1972.

[21]U.S. Atomic Energy Commission, *Environmental Statement, Rio Blanco Gas Stimulation Project* (WASH-1519), April 1972, p. 5-12.

[22]G. Holzer and D. O. Emerson, *Possible effects of the Rio Blanco Project on the overlying oil shale and mineral deposits*, Report UCRL-51163, Lawrence Livermore Laboratory Report, Dec. 1971.

[23]AEC, *Environmental Statement, Rio Blanco* (April 1972), pp. 5-11.

[24]Testimony of Dr. Robert Burnidge, New York University, in the informal public hearing on Project Rio Blanco, Denver, Colorado, March 27-28, 1972.

[25]*Ibid.*, p. 157. See also the *Draft Environmental Statement for the Proposed Prototype Oil Shale Lease Programs*, U.S. Department of the Interior, Sept. 1972.

[26]David J. Leeds and James R. Swaisgood, presentation before the Governor's Special Advisory Committee on the Rio Blanco Nuclear Stimulation Project, TOSCO, Appendix B, October 19, 1971, "Seismic Effects Prediction, Project Rio Blanco," Dames and Moore.

[27]Letters were sent by David Hunter of the Law, Technology and Politics Research Group staff to the following individuals: Mr. Herbert Holden, Jr., V.P., First National City Bank, New York; Mr. A. G. Gueymard, Senior V.P., First National City Bank, Houston; Mr. Leo Patterson, Jr., Senior V.P., First National Bank, Dallas; Mr. James K. Suhr, V.P., The First National Bank, Chicago; Mr. John A. Redding, V.P., Continental Bank, Chicago.

[28]A number of attempts by the Research Group staff to obtain a copy of Mr. Weickman's economic study met with failure. The Governor's Commission apparently did not obtain a copy.

[29]Civil Action No. C-36127, City and County of Denver District Court, 1973.

[30]*Rocky Mountain News*, May 2, 1973; see also the *Straight Creek Journal*, May 1, 1973.

[31]*Denver Post*, May 2, 1973.

[32]For detailed amplification of this argument and reasoning, see: Western Interstate Nuclear Board, *Plowshare Technology Assessment, Legal Studies*, Vol. I, 1973-74, pp. 1-2.

[33]*Ibid.*, p. 6.

[34]Colo. Rev. Stat., 1963, as amended, 66-28-1. The Water Pollution Control Commission was created under the police powers of the State to protect the health, peace, safety, and general welfare of the people of the State.

[35]Civil Action No. C-36127, p. 7.

[36]*Ibid.*, pp. 8-12. See also C. R. S., 1963, as amended, 66-28-1, - 2, 66-28-4, - 9, 66-28-11.

[37]Civil Action No. C-36127, p. 12.

[38]Colo. Rev. Stat., 1963, as amended, Section 3-16-5(7).

[39]Civil Action No. C-36127, pp. 17-21. All further discussion of the case refers to this transcript.

[40]Reported in *Boulder Daily Camera*, May 15, 1973, and in an interview with Mr. David Engdahl, June 12, 1973, by Catherine Wrenn.

[41]*Boulder Daily Camera*, May 15, 1973.

[42]According to the *Denver Post* (May 11, 16 and 17, 1973) and the *Rocky Mountain News* (May 16 and 17, 1973), the 10 a.m. time was originally chosen to allow local ranchers time to complete their morning chores before detonation and to allow them to resume their work after lunch. No explanation was given for moving the detonation up two days in the time schedule.

[43]*Rocky Mountain News*, May 18, 1973.

[44]*Ibid.* A series of earthquakes which rocked the Eastern Slope Denver metropolitan area during the 1950s and 1960s have since been traced to underground storage of water emanating from the U.S. Army Rocky Mountain Arsenal. This method of subterranean disposal has since been discontinued.

[45]*Rocky Mountain News*, May 18, 1973.

[46]*Rio Blanco News* [the official publication of Project Rio Blanco], August 1973. The information which follows on preliminary results through June 16, 1973, is drawn from this publication.

[47]*Rio Blanco News*, Jan. 1974.

[48]*Rocky Mountain News*, May 18, 1973.

[49]*Boulder Daily Camera*, July 12, 1973.

[50]*Denver Post*, April 29, 1973.

[51]*Ibid.*, and *Rocky Mountain News*, May 13, 1973.

[52]Dick Prouty, "Man and His World," *Denver Post*, May 13, 1973.

[53]*Denver Post*, June 27, 1973.

[54]*Boulder Daily Camera*, June 20, 1973. See also the *Denver Post*, June 20, 1973.

[55]*Rio Blanco News*, July 1974.

[56]*Denver Post*, April 18, 1974.

[57]*Ibid.*

[58]*Boulder Daily Camera*, June 11, 1974. Senator Haskell cited these figures in support of his effort to suspend continued research in similar nuclear stimulation experiments.

[59]*Denver Post*, June 20, 1974.

[60]*Denver Post*, Oct. 16, 1974.

[61]*Ibid.*

[62]*Denver Post*, Nov. 6, 1974. The immediate preceeding information on cesium-137 is taken from this article.

[63]*Denver Post*, Nov. 21, 1974.

Chapter
7
The Nuclear Impact:
Reflections Past and Present

The test of a good critic is whether he knows when and how
to believe insufficient evidence.

Samuel Butler

In the wake of two underground nuclear experiments in their
state, the voters of Colorado went to the polls in November 1974
and made it unequivocally clear that they would permit no
further nuclear stimulation of natural gas without their prior
consent. Soon after the elections, the industrial sponsors of one of
the experiments, influenced either by the vote of the people or by
the disappointing results of their experiments, or maybe a
combination of both, announced that they would not attempt to
sell any of the gas produced by nuclear means. Thus the people of
the state accomplished what neither the federal government, the
courts, nor the state and its governor were willing or able to do:
to halt further underground nuclear blasting in the state.

This development illustrates a trend which may be significant for
the future of the American political system. There seems to be a
growing power on the part of a politically and environmentally
conscious middle class which not only participates in the political
arena, but can also influence the course of economic affairs. In
Colorado, as elsewhere in the country, people are concerned
about the social costs resulting from the application of new
technologies and act accordingly. The term social cost, although
difficult to define or quantify because it depends on the value

209

judgment of the beholder, seems nevertheless applicable to the decision made by Colorado voters. It was their judgment that the possibility of long-term physical damage caused by continued underground nuclear blasting and the radiation hazards associated with this technology were too high a price to pay for a short-term increase in the supply of natural gas.

We believe that the people's decision to halt the nuclear stimulation of natural gas in Colorado was justified because (a) underground nuclear stimulation failed to produce a large and sustained increase in the flow of natural gas, (b) the economic advantage of using a nuclear rather than a conventional stimulation technology is questionable, and (c) there exists a possible health hazard from the radiation effects inherent in the production and continuous use of gas stimulated by nuclear means.

Although the voters' ability to halt further underground nuclear stimulation experiments in Colorado appears to be a vindication of the democratic process, an examination of the steps leading to the use of nuclear stimulation technology in the first place indicates serious shortcomings in the decision process which deserve further study in order to avoid making similar mistakes in other large resource-development programs in the future.

The potential dichotomy between the need for rapid energy development on the one hand and conservation of natural resources and preservation of the environment on the other is more complex now than it was in the past when a state or a community simply turned over its inventory of resources to the type of exploitation which appeared to offer the greatest financial reward. Complicating the picture of resource development is the fact that the issues raised relate to commodities or resources for which the free market place has elicited no realistic price information. Given this situation, some changes in political institutions are necessary to provide guidance to the technological efforts being made to stimulate large-scale resource development. The lessons learned from the Plowshare experiments in Colorado could well indicate the way to a more efficient and socially acceptable way of developing natural resources without undue harm to people and their environment. This development, however, can only proceed effectively if information is freely available and ample opportunity is provided to evaluate the positive and negative aspects of the technology in question. Sufficient time must be allotted so that the public can participate in the decision-making process before a point of no return has been reached. The implementation of this suggestion requires a technology assessment, a complex process which involves not only a technical evaluation of a proposed action but also an interrelated web of value judgments and political decisions.

In the broadest sense, we face today the problem of adapting modern technology to the requisites of the democratic process. If we rush into large-scale use of new technologies in order to fulfill our energy requirements today, unwanted side effects may develop and become difficult, if not impossible, to overcome tomorrow. To minimize the possibility of creating such effects, either general public agreement must be reached after complete information on the effects of the technology is available, or the government must act independently as the sole responsible authority. Neither alternative is ideal. The first runs the risk of unduly prolonging argumentation, as opinion becomes polarized on an emotional basis rather than on knowledge. The technology in question could meanwhile become so deeply entrenched that meaningful change might be difficult to institute. The second alternative can lead to a dictatorial technocracy which pays little heed to human values and can be a first step toward a political dictatorship of the right or the left.

We believe that whatever resolution is sought, it must lie somewhere between the two extremes, but must necessarily involve public education and opinion at the earliest possible time in the decision-making process. We recognize that there are inherent difficulties in taking this path. Since all results of a technological innovation cannot be foreseen and quantified, meaningful public discussion can be difficult. Contrary to public opinion, quantitative technical predictions are not inherently objective because the selection of relevant data and their interpretation involve some form of value judgment. Moreover, the price tag of human values is often not quantifiable in monetary terms, and the values chosen are usually not uniformly shared. Since it is not always clear what the precise adverse effects of a new technology may be, preventive action may require continuous monitoring and evaluation in the light of new information.

211

Although these and perhaps other difficulties exist, and participation in the democratic process is unwieldy and does not necessarily produce infallible results, it is feasible within the framework of state institutions to set up techniques which can facilitate the manner in which questions of public policy and technological innovation are handled. To be successful, these techniques must involve both the people and their elected representatives in the decision-making process. Although the exact political and administrative arrangements to accomplish this end may vary from case to case, the study of the use of nuclear stimulation technology in Colorado has led to some generalized conclusions which can help Colorado decision-makers faced with the implementation of other energy development programs—for example, strip mining of coal, the production of geothermal energy, coal gasification, or the large-scale processing of oil-shale deposits. Each of these programs has its unique technical problems, but they all share common social, political, and human problems which a democractic society cannot afford to ignore.

PLOWSHARE IN COLORADO: A REEXAMINATION

This study of Plowshare's underground nuclear stimulation projects provides some insights into the difficulties and dangers of scientific predictions. Technical predictions made by scientists and engineers differ, because in the process of extrapolating the results of an experiment to a large-scale application, the real system must be simplified, and this requires value judgments and choices. The decision-maker on the political level is therefore often faced with contradictory opinions and predictions by well-qualified technical experts. The underground stimulation program is a case in point.

In 1972, the AEC scientists and those from industry who cooperated in preparing the Rio Blanco Environmental Impact Statement claimed that "with the exception of the gaseous radioactivity discussed (i.e., tritium, krypton-85, carbon-14, argon-37, and argon-39 released to the atmosphere during flaring) it is not expected that any of the remaining radioactivity produced by the Project Rio Blanco detonation will be transported outside of the immediate cavity area. Most of this radioactivity is nonvolatile and will be permanently incorporated

212

in three zones of re-solidified molten rock (puddle glass) and the chimney region." On reading these assertions, the average citizen would have no reason to fear radioactive ground-water contamination from the Rio Blanco Project. But there were well-qualified members in the scientific community of Colorado who challenged this statement and its conclusion, and they questioned whether there was sufficient evidence at the time to sustain it.

Among the reputable scientists who questioned the validity of the claim was Dr. Edward A. Martell, President of the Colorado Committee on Environmental Information. In a letter to Congressman Donald Brotzman (now Undersecretary of the Army) dated March 27, 1973, Dr. Martell stated that in his view, the preceding quotation from the Environmental Impact Statement was misleading because it was "*well known* [italics added] that many radioisotopes produced in underground nuclear explosions have gaseous and volatile precursors and thus are not trapped in resolidified material." Dr. Martell went on to point out that the 90 kilotons of fission for Rio Blanco was expected to produce 17,000 curies of cesium-137 (28-year half life) and 10,000 curies of strontium-90 (28-year half life). He predicted that "most of the cesium-137 and about half of the strontium-90 would be present in water soluble form outside the resolidified rock, free to percolate through underground waters."

Dr. Martell further questioned the claim that other radioactivities will be *permanently incorporated* in resolidified rock, because, in his opinion, this prediction was not supported by experimental evidence. In view of the massive release of cesium-137, strontium-90, and tritium in the cavity volume in chemical forms which permit their transport through ground water percolating through the cavity, he contended that it should be incumbent upon the AEC and its industrial partners to provide detailed and convincing evidence that there would be no such transport on a time scale of decades or that the consequences would not involve unacceptable risks.

We are confronted here with a situation in which the predictions of a scientist not associated with the AEC and its industrial partners have apparently turned out to be correct, while those of the project sponsors now seem to have been incorrect. The claim of the impact statement apparently represented the technical judgment of some scientists in government and industry who were dedicated to the completion of a specific project. It is human nature to be enthusiastic about a project in which one is engaged, and it is therefore not surprising that in the process of selecting evidence there will be a bias in favor of the desired outcome. Therefore, to avoid one-sided technical judgments, we recommend that in the future the group of experts making predictions about the large-scale use of a new technology be enlarged to include some independent scientists who can freely examine available evidence on which decisions are to be based and present their conclusions without endangering their jobs.

In addition to the difficulty in evaluating technical predictions for large-scale projects involving natural resources, there is a problem in the evaluation of a project at intermediate stages of development. This problem is exacerbated by the pressures of publicity, particularly if the project has high visibility, as was the case with the underground detonations stimulation device. With all eyes centered on ground zero, one must guard against making preliminary results look better than the facts actually warrant. The initial official reaction to all underground nuclear experiments was that they had been a "success," probably because the nuclear devices detonated as expected without venting radioactive material into the atmosphere. However, a careful analysis of the final report on Gasbuggy issued by the Lawrence Livermore Laboratory two years after the detonation shows that the results of the project were disappointing because the expected permeability increase did not occur.[1] The objective of nuclear stimulation is to open pores and fissures in the surrounding geological structures containing natural gas so that increased permeability can facilitate the movement of the gas and augment the flow rate through the well. The AEC report indicates that the permeability increase resulting from the Gasbuggy blast in the San Juan Basin was restricted to a region within a cavity radius. By contrast, the Hardhat test, a previous underground nuclear shot, had led AEC scientists to expect a large permeability increase extending to four or five cavity radii. However, the Hardhat test had been carried out in granite, whereas the San Juan Basin is in a sandstone formation, and the permeability did not increase as significantly as had been expected from the Hardhat test. The Gasbuggy results should therefore have been a warning of problems facing tests in other gas-bearing sandstone formations. Although this information was available in a technical report, it was never mentioned in news releases by the AEC. Realism in the assessment of a new technology demands careful weighing of negative as well as positive factors, but objectivity of this sort is difficult to achieve when an agency depends for its continued funding on positive results of its activities.

Integral to the problem of balanced analysis and objective decision-making is the need for unimpeded access to sources of information. On various occasions, requests by scientists and engineers for information from the AEC were denied because the sought-after information was classified. It was therefore impossible even for a highly qualified scientist to make a thorough, independent analysis when he was not privy to AEC classified materials. At the same time, objections raised by those who questioned the Rulison or Rio Blanco proposals were often dismissed by AEC scientists who claimed that these objections were based on erroneous or incomplete information. There is no doubt that some nuclear information must be kept secret to protect our national security from external foes—but there exists also a need for internal security which demands that the widespread application of nuclear devices be postponed until it is possible to make a full and independent analysis of this potentially dangerous technology.

Without complete access to experimental data and other technical information, it is impossible to make a thorough assessment of tests already undertaken and to properly evaluate proposals for further experiments. As pointed out in connection with Project Gasbuggy, the unclassified analysis of the project's results was insufficient to predict realistically what might be expected in the Rulison test. To improve the stimulation intensity and the quality of the gas in the Rulison experiment, a different nuclear device was used by the AEC and its industrial partners, but the differences between the Gasbuggy and the Rulison devices were never made public. The published Gasbuggy test results appear to have been taken selectively, and comparisons were drawn with data from *some* conventional wells, but not from others. Because the AEC's conclusions were not based on all available data, they became questionable in the eyes of other scientists. At no time during the Gasbuggy or Rulison experiments were comparative tests of gas production made with a conventional well or a well stimulated by hydraulic fracturing at

215

the same depth and in the same geologic formation as the nuclear well. Only *after* the Rio Blanco test indicated that interchimney communication failed to occur did the sponsors of the project undertake conventional hydraulic fracturing experiments adjacent to and at the same depth as the Rio Blanco well. According to the sponsors of the hydraulic fracturing tests, the purpose of these new experiments is "to allow a comparison of hydraulic fracturing and nuclear stimulation for the recovery of natural gas from the tight impermeable formations that are prevalent in the Rocky Mountain states. . . . The gas production testing has been designed to allow a direct comparison between the two technologies."[2] It is difficult to understand why this type of controlled, comparative experimentation was not carried out much sooner in the overall design of Plowshare nuclear stimulation projects. If such a course had been followed, there would have been from the outset evidence on which to judge the viability of conventional alternatives to nuclear stimulation technology.

Any analysis of the viability of a new technology should also include some economic projections about the ultimate commercial value of the project that is to be produced. Although the Gasbuggy project was not designed to provide specific economic data, it did provide clues on the lack of economic potential of nuclear stimulation techniques. As noted previously, permeability increases were less than expected, the gas produced was highly radioactive, and its heating value was less than that of gas produced by conventional methods.

The Rulison and Rio Blanco tests, by contrast, were designed to produce economic yardstick data, and ultimately to produce commercially marketable gas. Dr. John Toman of the Lawrence Livermore Laboratory, speaking to the Tri-State Fossil Fuels Energy Conference in the winter of 1974, claimed that gas stimulated by nuclear means in the Piceance Creek Basin would cost approximately two-thirds of what Louisiana gas [presently] costs. He acknowledged, however, that a large number of detonations would be required to achieve this result. Neither the history of Rulison nor of Rio Blanco at the time seemed to bear out this optimistic conclusion, even considering the recently allowed fixed wellhead price of natural gas of 50 cents per thousand cubic feet. There is, of course, no precise way of predicting with accuracy the economic potential of a new technology in its testing stages. Yet the information released to the news media preceding each Plowshare test unfailingly implied that the nuclear stimulation method could ultimately produce large amounts of gas which would be economically competitive, or even cheaper, than conventionally produced gas.

The Austral Oil Company and the Colorado Interstate Gas Corporation spent approximately 11 million dollars on Project Rulison, but under today's price structure, no more than one-seventh of that amount could be recovered by the sale of the gas produced. Although the Rulison commercial backers were certainly counting on more nuclear shots and wells to produce a large volume of gas, the economic picture emerging from the Rulison well was certainly clear prior to writing the Rio Blanco Environmental Impact Statement in which a viable commercial enterprise was anticipated. An even more disturbing analysis was issued in April 1974 by the General Accounting Office (GAO). This report, requested by Senator Gale McGee of Wyoming after the Rio Blanco detonation, indicated that the Bureau of Mines had discovered evidence that the underground fractures caused by the Gasbuggy and Rulison detonations were beginning to close. Although the evidence is disputed by the AEC, the GAO report said that "if true, the wellhead cost of gas increases significantly depending on how quickly the fracture closes."[3] The study went on to detail the "uncertainties" involved in the nuclear stimulation process and concluded that "this technology is more expensive and poses more problems than massive hydraulic fracturing techniques." The report also noted that "as of June 30, 1973, there were no federal programs or funds for developing massive hydraulic fracturing," although the Bureau of Mines and the AEC budgets for fiscal 1974 included $1.5 million which could have been used for this development. Conventional mining is now underway on an experimental basis in the Piceance Creek Basin, but this action might well have come sooner if the technical data and economic evaluation of the Gasbuggy and Rulison experiments had been included in a technology assessment.

The final economic demise of the Rulison experiment became an acknowledged fact with the announcement by the President of Austral Oil, Miles Reynolds, that his company will not attempt to market Rulison gas. According to Mr. Reynolds, "the timing is bad and we'd have a difficult time in light of opposition to nuclear matters and the difficulties at Rio Blanco. So we're just sitting tight."[4]

Actually, the AEC industrial sponsor did not simply "sit tight," but attempted to recoup a part of its investment without selling Rulison gas. In 1974, the Austral Oil Company sued the AEC to recover $5.82 million in costs for the project because, the company claimed, "it relied on the government's research on potential recovery of natural gas in planning the experiment, but only a small amount of gas was produced."[5]

Economics, consumer resistance, and technical failure marked the end, at least for the foreseeable future, of any attempt to market gas produced by nuclear stimulation. The term "consumer resistance" perhaps illustrates as clearly as any other an added dimension and problem in the effort to develop and use underground nuclear technology for peaceful uses. The decision to use a new technology must finally rest not only with the scientists who have devised the technique and estimated its economic potential, but also with the people who will ultimately reap its rewards or liabilities. In considering how and by whom such decisions should be made, there are at least three avenues of approach.

First, a new technology may be used without any specific regard for ultimate social or environmental costs. The "invisible hand of the market place," the law of supply and demand, and the ability of a technological innovation to generate the capital necessary for its application are all that is required. However, this approach has been all but foreclosed with the passage of the National Environmental Policy Act in 1969 and a growing public and official awareness of the limits of the planet Earth, including its atmosphere.

Secondly, an alternative approach to that of unregulated free enterprise has been advocated by the economist Glenn Hueckel,[6] who suggested that the use of a new technology which could involve substantial social costs and side effects should not be settled by the consumer alone. He suggested that the individuals, whether from government or private enterprise or both, who promote a technology should be forced to pay its full social costs. It then becomes the responsibility of policy-makers to see to it that appropriate payments be made through the use of charges established by regulatory controls. Such a policy would, of

218

course, raise the cost of the produced commodity and transfer the damaging side effects from society at large to those initiating the technology and producing the commodity. Presumably such tough regulatory controls would not preclude damaging effects from the use of a new technology altogether, but they would make them subject to the law of the market place. If including the social costs in the initial production costs should raise prices beyond the point that the market can bear, there would be an inhibiting effect on the enterprise which would reduce or even eliminate the damaging effects.

The third approach involves direct evaluation of a new technology and its potential social costs by the people and their governmental representatives before its implementation. This course can prevent potential adverse effects and social costs of a technology before they occur by simply preventing its applications. This in fact was the course taken by Colorado voters in passing Amendment 10.

Whatever approach is taken, however, the action will inevitably occur in the political arena where technology, law, and politics interact and where human and economic values must be balanced.

The importance of regulatory controls in the development of a new technology have already been alluded to. This history of the Plowshare Program illustrates not only the role of the AEC as the prime promotor of nuclear technology for the recovery of natural gas but also its companion role as chief regulator of the associated radiation effects. In essence, the AEC could prevent the commercial use of nuclear-stimulated gas by not issuing the necessary marketing license, but it could equally well push forward with one nuclear experiment after another without setting radiation standards. The dual role of one agency being responsible at one and the same time for the development and the control of a potentially hazardous technology has been widely criticized and discussed in connection with the development of nuclear power plants. The remedy to the conflict of interest in promoting and controlling nuclear Plowshare technology is, of course, to separate these two functions. This step was actually taken before this study was completed. The problem was resolved by dividing the promotional and regulatory functions previously held by the AEC between the new Energy Resource Development Agency (ERDA) and the Nuclear Regulatory Commission.

The question of who has ultimate regulatory authority inevitably brings federal-state relationships into play. During the course of the Gasbuggy and Rulison projects, there was little evidence of federal-state disagreement or conflicting interests regarding the nuclear detonations. The Rio Blanco project, however, was a different matter; over a period of many months, it almost produced a federal-state conflict. Given the current national energy problem, it is obvious that overall national energy policy must be coordinated and based on carefully determined priorities. It is equally clear that the states and the citizens in the states affected by the implementation of this policy want to have some control over the development and utilization of their land and resources. The history of Rio Blanco indicates not only a lack of federal policy in thoroughly evaluating alternatives to nuclear stimulation and the compatability of that technology with future oil-shale development, but also a breakdown in communication with the Governor of Colorado. As a result, the question of how to conduct underground nuclear explosions within the state was approached incrementally and without an overall policy.

Until six months before the Rio Blanco devices were detonated, it was widely believed by citizens in Colorado that their Governor could veto the project. All the procedures for implementing the Rio Blanco plan thus seemed to drift along, while state agencies as well as citizens were under the impression that the Governor would have the final word. Although some state agencies had participated in some of the preliminary planning of the project, their participation was piecemeal, and it was not evident that any of them ever considered the overall design or the long-range implications of the experiment. When it became apparent that the Governor would not or could not halt the project, a number of state agencies were called upon to carry out their missions within the framework of the state Constitution and legislative requirements. In the absence of an agency with overall responsibility, several agencies were forced to act alone and incrementally. One agency granted a use permit for the land under which the nuclear detonation was to take place, another permitted the flaring of radioactive gas into the atmosphere, and yet another issued a permit for the disposal of radioactive water produced by the nuclear blast. The action of the state Water

220

Pollution Control Commission and its successor, the Water Quality Control Commission, in the latter instance was particularly illustrative of this incremental approach. After the hearing in which the Commission accepted the AEC claims that there would be no radioactive pollution of ground water, it permitted the project to proceed. The acting chairman announced that consideration would be given to possible methods for the disposal of radioactive water *when the occasion arose.* When the occasion did arise, radioactive waste was a *fait accompli* and had to be handled within a limited range of options. During the decision process, one agency action followed another, and at no time was there a broad overview of the project and its attendant ramifications by one specific state agency prior to issuance of the final permit to allow detonation.

There are no easy answers to the question of federal-state relationships, particularly when a clear-cut national energy development policy is at stake. As the voters of Colorado demonstrated in passing Amendment 10, however, the people most directly affected by the policy want to be consulted about the ways their land is to be used. They want to know clearly and publicly the technical basis for a decision to proceed with a new technology. Such presentation should first of all contain a comprehensive evaluation of alternatives, including an analysis of the economic, social and environmental costs and benefits. Secondly, a procedure for direct consultation with the states directly concerned with the use of a new technology should be set up. In the future, there must be a clear delineation of procedure and authority in these procedures, or the people will increasingly feel compelled to take the decision-making process upon themselves through initiative and referendum.

During the course of the Plowshare underground nuclear stimulation program, the courts were asked on several occasions to resolve controversies regarding the regulatory authority of different agencies as well as to rule on technical issues such as the acceptable level of radioactive contamination resulting from the flaring of gas. Since it is likely that the courts will again be called upon to provide guidance, if not resolution, in controversies arising from other resource developments, the place and proper role of the judiciary in such matters should be reexamined. We believe that the courts should settle questions related to federal and state rights, the regulatory authority of agencies, and compliance with safety standards, but they should not be called on to settle technical controversies. In technical questions such as the safe level of radioactivity resulting from a nuclear operation, the burden of proof has rested on the plaintiff, who, in the past, often had only limited access to information and whose financial resources were limited. It was the plaintiff's responsibility to demonstrate to the court that the use of the proposed technology was hazardous and would in the long run be detrimental to human health and environmental quality. Judges are in general not trained to rule on the relative merits of scientific arguments and therefore tend to rely on such criteria as the credibility of the witnesses. Since a large governmental agency or an industrial sponsor has usually greater financial resources, better access to information, and a readily available pool of experts at its disposal, the scale of justice may not be evenly balanced. The problem is compounded for a judge because of the piecemeal and incremental nature of the decision-making process in launching a new technology. As a result, litigation often occurs at a late hour and in the political limelight. At that point, when all engineering preparations have been completed and the required steps in the systematic decision process have been taken, it becomes difficult to halt the flow of anticipated events, even if public opinion should not be in favor of the outcome. We believe that in the future the courts should not be asked to make an essentially technical decision at the last moment under such circumstances.

It is in some respects tempting to conclude that because of the apparent failure of the Rio Blanco project, the problems posed by the various attempts to bring the Plowshare Program to commercial life are now no longer relevant; but in spite of apparent industrial disenchantment and public concern, there is no assurance that nuclear stimulation technology is dead. The federal government may decide to continue experimentation alone, if not in nuclear gas stimulation, then perhaps in oil-shale development. As late as March 1974, a House of Representatives Mining Subcommittee was told by a Bureau of Mines official that the nuclear approach to oil-shale mining would avoid the problem of disposal of spent shale above ground.[7] This official testified that the Lawrence Livermore Laboratory had updated earlier nuclear oil-shale development plans but that no field test was scheduled at the current time. Dr. Edward Flemming, the Acting Director of the AEC's division of applied technology,

222

said, in commenting on the apparent Rio Blanco failure and waning industrial support for the Plowshare Program, that "the Government could go it alone. I'm not predicting that it's going to happen that way, but it's always a possibility."[9]

Those and other such possibilities make relevant the implications of the recurring themes that have become apparent in the exploration of the Plowshare Program. It may well be that the energy crisis which engulfs the nation with increasing momentum will dictate a course which includes further underground nuclear experimentation and development, but recent history may provide some clues to a more prudent and efficient course than has been followed in the past.

TECHNOLOGY ASSESSMENT AS A GUIDE TO FUTURE ACTION

It is generally recognized that the federal government participates in large-scale projects through actions such as regulations of interstate commerce, tax incentives, and subsidies for research and development. These governmental actions inextricably intermesh technology and public policy in our society; in the past decade, governmental decisions in technologically oriented matters have become increasingly significant in our lives. Moreover, as the scope and pace of technological innovation increases and the lead time between new ideas and their implementation decreases, these decisions involve larger and larger risks. To cope with these risks and at the same time to reap the benefits of increasing technical knowledge without infringing individual options and liberties is the challenge of our technological age. In our study, the problems in setting satisfactory radiation standards illustrate this trend.

A state can make its influence felt in federal energy development decision making only to the extent that its own policy is authoritatively defined and unequivocally presented at the national level. There is no guarantee that the federal government will be bound by the dictum of the people of the state that they be given the opportunity to allow or disallow further nuclear detonations. However, a vocal citizenry clearly has had an effect on the industrial partners of the federal government in Plowshare projects in Colorado. These cosponsors now seem hesitant to proceed with any more nuclear experimentation. The influence of the state and its people is thus likely to be persuasive, if not compelling, in the determination of national policy.

In this situation, it is all the more important for Colorado to develop procedures and policies regarding the use of a new technology in a manner that is comprehensive and informative to the citizens of the state. Since the state is now bound to respect the wishes of its citizens in regard to further underground nuclear blasting, the state should also present clearly the issues involved, any gaps and discrepancies in scientific opinions, and information about possible alternatives.

As the history of the Plowshare Program in Colorado illustrates, there exists as yet no comprehensive overview of the implications of large-scale nuclear stimulation or other resource-development programs, and state commissions and boards can therefore only react as particular issues arise.

Although there are certainly no easy solutions to the administrative, political, legal, and technical ramifications of a new technology designed to exploit natural resources within a state, the state government should at least provide sources of information about them. The recommendations for public information procedures and a decision-making process proposed here are principally directed to the State of Colorado, but with minor variations, they could be adapted to other states planning large-scale resource development.

The core of our recommendations centers on a technology-assessment process carried out at the state level. The basic purpose of technology assessment should be to improve and make more rational the means by which key public policy decions are made in a democratic society. The assessment process is particularly important when large-scale governmental actions are taken which allocate public resources to stimulate or control the development of a new technology.

The end result of a technology-assessment program should be to encourage developments which are socially, economically, and environmentally in the public interest. It could be applied, for example, to projects such as mass transit systems and pollution control, as well as to large-scale development of natural resources. For projects in which detrimental side effects seem inevitable, the assessment process should investigate different options for achieving the project goal and consider the costs and benefits of each option. On this basis, pollution controls, licensing arrangements, land reconstitution, and other methods to ameliorate detrimental impacts should be recommended. In cases in which future impacts are uncertain and impossible to evaluate at the time a decision must be made, but the project appears to be in the public interest, the process should propose careful monitoring of all side effects after such a technology has been implemented. If, after careful analysis, a project is deemed excessively dangerous and cannot be justified in terms of public interest and welfare, this information should be passed on to the appropriate decision makers for action.

The technical assessment process should be a routine part of decision making at the state level, but there should also be opportunity for an external check on the government by independent experts acting as private citizens or representatives of interest groups. There should be at least three mechanisms by which citizens and their elected representatives can play a part in decisions to allow or disallow the use of a new technology.

First, a group of ten or more state legislators should be entitled to request a technical assessment when they consider it necessary.

Second, the Governor of the state, in consultation with the Director of Natural Resources, should also be able to request such an assessment.

Third, the citizens of the state should be permitted to request an assessment through the initiative and referendum. This was the procedure used by the voters of Colorado to block further underground nuclear blasting in the state.

The key objective of a technology assessment is to provide as unbiased an analysis as possible and consider all available data and options for the use of the decision-maker who must ultimately weigh the trade-offs between benefits and costs. The objectivity of the assessment is clearly of paramount importance if bias toward one choice or another is to be prevented. It should, however, be fully recognized that no evolving technology is free of value judgment and bias, because there are always some who expect to benefit and some whose interests may be hurt. In particular, there is often a sector in government or business which looks forward to an immediate gain from the implementation of a new technology. Since this sector is usually best informed about the technology and has the most at stake, it can often do the most effective job of influencing public opinion. To guard against the bias of those who anticipate immediate gain is a principal task of a technology assessment, and this can be achieved only by soliciting a diversity of informed analyses from qualified experts who have nothing to gain or lose as a result of their judgments.

Since, as we have noted before, it is almost impossible to obtain a truly objective group of people, it is necessary to select qualified experts with different scientific, political, and economic views for the assessment team. The mechanics of the selection process should be procedurally established in order to include as broad a spectrum of opinion and expertise as possible. A report on existing mechanisms for state technical assistance programs has recently been compiled at the University of Washington with suggestions for improvements which seem warranted on the basis of experience to date. There is every reason to believe that with more experience, a satisfactory procedure for selecting members of assessment teams can be developed.[10]

Although the Governor's Commission on Rio Blanco described in Chapter 5 may appear in some respects similar to a technology assessment team, in practice it was a vastly different enterprise from what is recommended here. In the first place, the charge to the Commission was narrow because there was no provision for a survey of alternatives or options; the question was simply whether to proceed with the Rio Blanco stimulation. In addition, there were no pre-established ground rules for the selection of Commission members or specific procedures for public involvement. The report of the Commission was not made public for a number of months after the Governor received it, and the majority and minority versions did not receive similar publicity. Finally, the Governor never cited specific data on both sides of the issue which guided him as the decision-maker. In short, the Governor's Commission went to work in an *ad hoc* effort, and because it began operation late in the total developmental scheme after many millions of dollars had already been spent, it had no realistic chance to change the course of the project. In fact, it may be that the power of decision of whether to proceed with Rio Blanco was out of the Governor's hands at the time the Commission report was completed; the federal government may have by that time decided to proceed, regardless of the Governor's wishes. However, a thorough technology assessment program, initiated at the state level at an early stage of the development of a new technology, could have provided the Governor and the people with expert, factual information on which to base a position, and the Governor would have had a solid basis for reaching a decision, a fact which the federal government could ill afford to ignore.

Until recently, the only sustained source of funding for comprehensive technology assessments has been on the federal level through the National Science Foundation in its program on Research Applied to National Needs (RANN). Two years ago, however, the Office of Technology Assessment (OTA) was established by law and began operating in 1974 as the applied science research and assessment arm of Congress. During its first year, OTA has concentrated on such topics as transportation, biochemicals, food, oceans, energy, and mineral sources. At the present time, rudimentary technology-assessment groups are also emerging on the state level. For example, the governors of Utah and Hawaii have established an Office of Science Advisor with a technology-assessment capacity. In Colorado, it appears that a capability for making technology assessments should be located in the Department of Natural Resources, because proposed large-scale uses of any new technology will have their most immediate impact on the natural resources of the state. We therefore recommend that a permanent Office of Technology Assessment be set up within this department so that it can have an established position within a functioning agency.

The Office of Technology Assessment should be responsible for a broad overview of large-scale implementation of new technologies and resource developments in the state without having to resort to the piecemeal decision process forced upon state agencies during the Rio Blanco project. However, active contributions by specialized agencies of the state government, such as the Departments of Health, Labor, and Education and the Water Quality Control Commission and the Oil and Gas Commission are imperative to the broad overview required by a technology assessment and should be made a part of the procedure. The schematic diagram in Figure 7-1 depicts the mechanism by which a technology assessment process could be initiated and carried out in the state of Colorado.

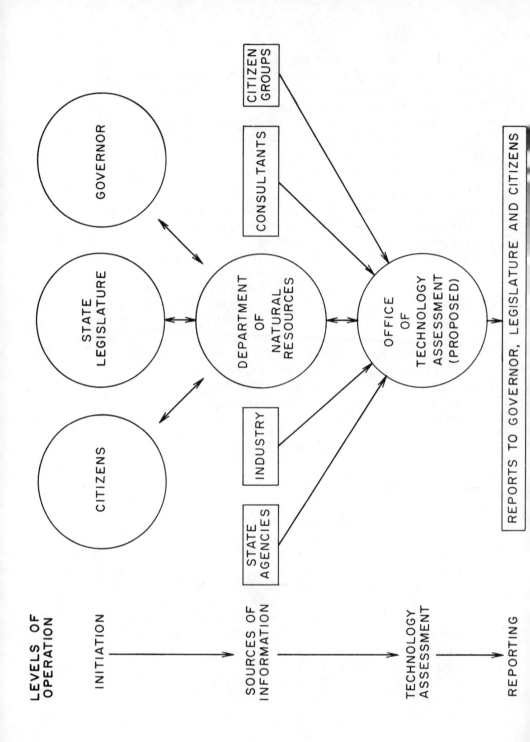

LEVELS OF OPERATION

INITIATION

SOURCES OF INFORMATION

TECHNOLOGY ASSESSMENT

REPORTING

CITIZENS

STATE LEGISLATURE

GOVERNOR

STATE AGENCIES

INDUSTRY

DEPARTMENT OF NATURAL RESOURCES

CONSULTANTS

CITIZEN GROUPS

OFFICE OF TECHNOLOGY ASSESSMENT (PROPOSED)

REPORTS TO GOVERNOR, LEGISLATURE AND CITIZENS

228

State funds should be provided for the core of a technology-assessment staff, but the experts involved in any specific assessment should be selected in accordance with the requirements of the task. All experts involved in an assessment should be paid on a regular basis, because specialists who are not attached to large-scale business or government can generally not afford to take on such a task. Although it is difficult to make cost estimates, it would seem that funds on the order of $200,000 per year would be sufficient to cover the functions of an Office of Technology Assessment in the state.

Perhaps the most compelling reason for establishing such an office is the opportunities it would open for public education and debate after an assessment has been made. A technology assessment is not a decision-making process, and it is not intended to take over the duties and prerogatives of elected decision makers. It is intended to provide information and to outline a range of available options. Public awareness of the immediate and future effects of a new technology is particularly important in the initial stages of development when problems are being defined and alternatives identified. If the public is not involved in the early stages of this process, it has lost the most important opportunity to influence the final decision. In the absence of public participation, the values of a particular interest group could well come to dominate and to severely limit the kind of options which emerge in the political arena. This can impair the ability of a democratic society to guide technological development toward its own best interest. Although technology-assessment procedures are new and imperfect, they can help to steer a better course for making public policy decisions in a democratic society.

Irrespective of the future of Plowshare, decisions on the implementation of new technologies and natural resource developments will require value judgments. We hope that in the future the people and their elected representatives will have the opportunity to make their judgments on the basis of a broad spectrum of information, differences of opinion, and potential alternatives. Whether we are concerned with nuclear or conventional technologies, the ultimate impact will be felt by the people and the Earth they inhabit. It is, after all, for the people and by the people that swords are beaten into plowshares and spears turned into pruning hooks.

NOTES

[1] O. Borg, "Microfracturing in Postshot Gasbuggy Core GB-3," *Nuclear Technology II*, July 1971.

[2] News release, CER Geonuclear Corp., Las Vegas, Nevada, March 27, 1974.

[3] *Boulder Daily Camera*, April 4, 1974.

[4] *Rocky Mountain News*, Dec. 11, 1974.

[5] *Boulder Daily Camera*, Nov. 14, 1974.

[6] Glen Hueckel, "A Historical Approach to Future Economic Growth," *Science*, March 14, 1975, p. 929.

[7] *Boulder Daily Camera*, March 15, 1974. See also Bill Jordan, "Scientists Unveil Blast Schemes," *Boulder Daily Camera*, Feb. 1, 1974. As explained by Dr. Arthur Leeves of the Geoscience Division of the Lawrence Livermore Laboratory, the nuclear method of recovery of oil-shale resources would involve a series of nuclear explosions, set off in vertical arrangement within the oil-shale layer, which would form a nuclear chimney. The shale could then be retorted within the chimney and piped to the surface. This method, he said, would produce oil at a much lower cost than other proposed methods ($3.50 a barrel as compared with $5.50 a barrel for open pit mined shale and $7.00 a barrel for underground mined shale). It would also eliminate the huge population build-up in the Piceance Creek Basin which is foreseen in connection with other proposed methods.

[8] *Boulder Daily Camera*, Dec. 16, 1974.

[9] J. F. Coates, "Technology Assessment and Public Wisdom," *Journal of the Washington Academy of Science, 65 (1975)*, pp. 3-12.

[10] Jim Craig *et. al., Technology Assessment and Technical Assistance for Washington State Legislators: An Advisory Model for State Government.* Programs in Social Management of Technology, June 1974, University of Washington, Seattle, Washington.

APPENDIX

Correspondence of Dr. E. A. Martell, President, Colorado Committee for Environmental Information, Boulder, Colorado, 80303, relative to Rulison and Rio Blanco nuclear gas stimulation experiments

10 January 1975

Dr. E. A. Martell
President, CCEI
325 Norton Street
Boulder, Colorado 80303

Professor Frank Kreith
Department of Chemical Engineering
College of Engineering and Applied Science
University of Colorado
Boulder, Colorado 80302

Dear Professor Kreith:

For your information I enclose a copy of the CCEI comments on the
Rio Blanco Project which I presented at the public hearings at the Federal
Center in Denver on March 27, 1972. I also enclose an earlier individual
statement I sent to the Colorado Environmental Commission, dated March 2,
1971, in which I first pointed out that most of the cesium-137 and strontium-90
will be present, outside the fused rock, in water soluble form.

There is no objection to your publication of either or both of these
statements, in part or in full, in the text or appendix of your forthcoming
publication.

Sincerely yours,

Edward A. Martell

Encls.: Statements of March 1971 and March 1972

March 2, 1971

Chairman
Colorado Environmental Commission
600 State Social Services Bldg.
1575 Sherman Street
Denver, Colorado 80203

Dear sir:

On Tuesday afternoon, February 9, 1971, I discussed the problem of plutonium contamination of the Denver area from the Dow Chemical Company Rocky Flats Plant with members of your commission. Later in the afternoon I also commented on the problem of radioactive contamination of underground water supplies by future Rulison type nuclear gas stimulation experiments. My comments and recommendations on the latter topic are enclosed.

Sincerely yours,

E. A. Martell

EAM:rer

Enclosure

<u>Radioactive Contamination from Rulison Type Nuclear Gas Stimulation Experiments</u>

In order to limit the extent of radioactive tritium contamination of the natural gas, the AEC Plowshare program uses 100 percent fission devices in nuclear gas stimulation experiments. This introduces very serious radio-active contamination risks in any future large scale application of this technology, including the following:

1. Accidental venting of the radioactive products into the lower atmosphere could produce extremely serious levels of fission products in local downwind areas, particularly iodine-131, strontium-89, strontium-90 and cesium-137. Such ventings are most likely to occur for the shallower underground explosions or via natural or explosion induced faults in the overlying rock structure.

2. Underground water contamination, particularly by strontium-90 (half-life, 28 years), cesium-137 (half-life, 28 years), and tritium (half-life, 12.6 years) will persist for many decades and may become widespread, depending on the rates of groundwater movement in the affected areas. (Rulison proponents have tried to minimize the possible significance of underground water contamination by stating that most of the radioactive products are locked in the glass formed by condensation of fused rock in the lower portion of the cavity, and by pointing out the very slow rate of groundwater movement experienced in the Nevada test area. However most of the long-lived strontium-90 and cesium-137, two of the fission products of greatest biological concern, will be dis-tributed outside the glass, throughout the cavity, in water soluble chemical form. Most of the radioactive tritium pro-duced will be present as tritiated water and therefore will mix and flow with underground waters without restriction. Furthermore, the slow underground water movement experienced in the dry subsurface levels of the Nevada test site is very unlikely to be applicable to Colorado western slope areas.)

3. Radioactive krypton-85 gas (half-life, 11 years) as well as radioactive tritium (as tritiated water) both are released into surface air in the flaring process and thus will con-tribute somewhat to radiation exposure of people in downwind areas. The tritiated water also will accumulate in downwind vegetation and surface waters. If the mixture of radioactive gases produced in the underground explosions is not subjected to flaring, the chemically inert radioactive krypton-85 will slowly seep up through faults and soils into surface air any-way. In addition, part of the tritium which is incorporated into the natural gas, as well as krypton-85 which is mixed with it, will be distributed with the gas. (Although none of these sources of radiation poses a very serious risk to the public, each will contribute an undesirable additional radia-tion exposure to local residents and to the natural gas users. Thus if there are competetive non-nuclear methods of stimu-lating these tight natural gas shales, and there probably are, it would be preferable to use them.)

234

4. Underground structures directly contaminated by radioactivity
   from nuclear blasts as well as those in surrounding areas contaminated by underground water perculating through them, contain mineral resources other than natural gas which may be
   of value. Depending on the levels of radioactive contamination
   involved, access to such mineral resources in contaminated areas
   underground may be restricted, or even denied, for many years to
   come.

Of the above mentioned possibilities, the problems of radioactive
contamination of underground waters of Colorado western slope areas is likely
to be the most serious consequence of large scale application of nuclear gas
stimulation technology. Detonation of 10,000 all-fission explosions, each of
40 kilotons energy release, would be involved in the substantial exploitation
of natural gas in western Colorado. This total of 400 megatons of fission
would introduce 40 million curies of strontium-90 and 70 million curies of
cesium-137 into subsurface waters. One hundred years later nearly 10 million
curies of long-lived fission products would still be present underground.

It should be noted that nuclear tests in Nevada, which have released
much smaller quantities of fission products underground, have rendered an area
of several hundred square miles unfit for human habitation and for few uses
other than nuclear testing for many decades to come (See USAEC report, Environmental Statement, Underground Nuclear Test Programs, Nevada Test Site, November 1970.) Areas that would be similarly affected by nuclear gas stimulation
technology include not only the underground explosion sites but very much
larger areas through which contaminated underground waters perculate for the
next 100 years, or more.

On the basis of the above considerations I make the following recommendations to the Colorado Environmental Commission:

1. That the acceptability of the large scale application of nuclear
   gas stimulation technology be critically reviewed at the state
   and national level, before further Rulison type experiments are
   authorized. Such review would best be carried out in open public
   hearings conducted by the Environmental Protection Agency or
   another appropriate federal agency other than the AEC.

2. That the reliable prediction of contaminated underground water
   movement for the next 100 years should be made a condition of
   approval for any future underground nuclear explosions in
   Colorado.

E. A. Martell

Dr. E. A. Martell
325 Norton Street
Boulder, Colorado   80303

March 27, 1972

Comments on the AEC Environmental Statement

on the Rio Blanco Gas Stimulation Project

    In January 1972 the United States Atomic Energy Commission issued a Draft Environmental Statement on the Rio Blanco Gas Stimulation Project, AEC Report WASH-1519. That statement was prepared in support of the AEC's participation in the proposed underground nuclear explosion in Rio Blanco County involving simultaneous detonation of three nuclear devices with an aggregate yield of about 90 kilotons.

    Members of CCEI have reviewed this AEC environmental statement and we find that the statement is misleading or inadequate in several important respects. It is in the public interest that these deficiencies be corrected in the AEC's final environmental statement. In addition we recommend that certain other steps be taken before the Rio Blanco Project or any other nuclear gas stimulation demonstration experiment is approved and conducted in the public sector. Our comments and recommendations follow:

1. Ground Water Contamination: The AEC's discussion of the possible extent of ground water contamination from the Rio Blanco Project is seriously incomplete and misleading when it is stated (p. 35):

> "With the exception of the gaseous radioactivity already discussed (i.e., tritium, krypton-85, carbon-14, argon-37 and argon-39 released to the atmosphere during flaring) it is not expected that any of the remaining radioactivity produced by the Project Rio Blanco detonation will be transported outside of the immediate cavity area. Most of this radioactivity is nonvolatile and will be permanently incorporated in three zones of resolidified molten rock (puddle glass) and the chimney region."

    This is very misleading because it is well known that many of the radioisotopes produced have gaseous and volatile precursors and thus are not trapped in the resolidified material. The 90 kilotons of fission for Rio Blanco is expected to produce 17,000 curies of cesium-137 (28 year half-life) and 10,000 curies of strontium-90 (also 28 year half-life). Most of the cesium-137 and about half of the strontium-90 will be present in water soluble form, outside the resolidified rock, free to perculate through under-

A non-profit corporation of scientists providing analytical reviews of public issues concerning the impact of technology on our environment.

ground waters. This is especially disturbing because cesium-137 and stron-
tium-90 are two of the most dangerous and biologically active radioisotopes
known.

The AEC claim that other radioactivities will be "permanently incorporated" in resolidified rock also is unsupported and questionable. It
is not established that this material cannot be leached by ground waters to
a substantial degree on a time scale of many decades. Finally, the report
is vague about the AEC's plans for the disposition of the nearly 3,000 curies
of tritiated water that is anticipated.

In view of the massive release of cesium-137, strontium-90 and
tritium in the cavity volume in chemical forms which permit their transport
by ground waters which may perculate through the cavity, it is incumbent upon
the AEC and its contractors to provide more detailed and convincing evidence
that there will be no such transport on a time scale of many decades or that
the consequences involve acceptable risks.

These considerations apply not only to the potential radioactive
contamination of local underground water resources but also to the question
of the nature and extent of contamination of oil shales and other mineral
resources. The pattern and rate of transport of contaminated water as well
as radioactive gases through natural and explosion induced faults should be
more fully understood before such experiments as Rio Blanco are undertaken.

2. Alternatives: Discussion of the alternative methods of developing the
Rio Blanco Field is too brief and superficial, especially in the light of
the risks involved in nuclear stimulation. The three alternatives, (1) nuclear stimulation, (2) conventional hydraulic fracturing, and (3) emerging
techniques involving ordinary high explosive stimulation, all should be
evaluated in detail with respect to gas yield per unit cost and including
all environmental costs and side effects [see NRDC, V. Morton (D.C.D.C.)
3 E.R.C. 1558 (1972)].

It appears to our committee that the development of nuclear gas
stimulation technology is proceeding in a vacuum, without adequate evaluation
of non-nuclear approaches and support of their development. This procedure
appears to us to represent a misuse of the taxpayers' money and points up
the pressing need for a federal power authority to insure a balanced and
responsible approach to the development of our fuel resources to satisfy our
national energy needs.

3. Benefits versus Risks: In their discussion of benefits and risks in
Section 10 and elsewhere in the statement, the AEC repeatedly points to the
benefits in gas production from the full development of the Rio Blanco field
and other fields but their discussion of risks is entirely restricted to the
possible side effects of the single Rio Blanco Project explosion. It is obvious that what is needed is an evaluation of the benefits and risks resulting from the large scale application of nuclear gas stimulation. What will
be the environmental impact of hundreds of nuclear explosions in the Rio

3

Blanco field and in the Rulison field? What will be the environmental impact
of several thousand nuclear explosions in the upper Colorado River basin?
Unless it can be shown that such large scale applications do not involve unacceptable risks to the environment and to the public, there can be no justification for the AEC to conduct the Rio Blanco shot or any other nuclear
stimulation experiment.

It must be added that any resonable assessment of the risks and
benefits of Rio Blanco should be made in the context of the fullest possible
disclosure and evaluation of the results of the Projects Gasbuggy and Rulison.
The environmental risks are being increased in each of a series of increasingly large nuclear gas stimulation explosions. Each successive experiment
can be rationally evaluated and justified only after full analysis of the consequences of previous experiments. Such essential disclosure and analysis
of past experience form no significant part of the present AEC statement.

4. Energy Waste: A comprehensive evaluation of alternative methods of gas
stimulation and the benefits and costs involved should include evaluation
of the energy consumed in the process for each method. The AEC statement
omits discussion of (1) the electrical energy consumed in the manufacture of
the nuclear explosive devices to be used in nuclear gas stimulation applications, (2) the energy represented by the fissionable material of these devices, were it to be used in fission reactors for electric power production,
(3) the natural gas consumed and wasted as the result of its dissociation
and oxidation in the cavity under the influence of the intense heat and
radiation from the explosion. The presence of a large excess of carbon dioxide, carbon monoxide and molecular hydrogen in nuclear explosion stimulated gas may be explained in large part by dissociation of methane by
ionizing radiation and chemical reactions of the dissociation products.

If nuclear gas stimulation consumes significant amounts of our
dwindling energy and fuel resources this fact should weigh heavily in the
cost assessment of alternative methods of gas stimulation. A quantitative
estimate of such energy consumption and fuel waste is essential to the
proper evaluation of nuclear gas stimulation versus alternatives. To have
omitted these vital estimates in the discussion of costs and alternatives
reduces this aspect of the AEC environmental statement to mere promotional
literature.

5. Radiation Exposure Standards: The question of what constitutes an acceptable level of exposure to the public in the course of flaring the radioactive
gas as well as in the distribution of radioactive gas to consumers has not
been adequately resolved and is hardly a matter which should be left to the
self-interested discretion of the agency which is promoting the technology.
There may be some justification for the inadvertent radiation exposure of
the general public in the course of the production and testing of nuclear
weapons in the national defense interest. However it is an entirely different matter for the AEC to decide what constitutes acceptable levels of public
exposures from the peaceful uses of nuclear explosives.

In connection with public radiation exposure from flaring operations
the AEC states, p. 27, that their position is based on that established for

radioactive consumer products, as follows:

> "Approval of a proposed consumer product will depend on both
> associated exposures of persons to radiation <u>and</u> <u>the</u> <u>apparent</u> <u>use</u>-
> <u>fulness</u> <u>of</u> <u>the</u> <u>product</u>. In general, risks of exposure to radiation
> will be considered acceptable if it is shown that....it is unlikely
> that individuals in the population will receive more than a small
> fraction, less than a few hundredths, of individual dose limits
> recommended by such groups as the ....ICRP, the ....NCRP, and the
> ....FRC" [emphasis added].

The AEC then states that "Presumably the same risk-benefit considerations would be applicable here" (to public exposures from flaring).
On the contrary, those exposed to additional radiation from Rulison and Rio
Blanco flaring operations are in most instances neither users nor beneficiaries of these AEC experiments.

It appears to us that such questions as acceptable radiation exposures to the public from nuclear gas stimulation experiments and other
AEC Plowshare projects should be specifically addressed by the NCRP, by the
Department of Health, Education and Welfare and by the Environmental Protection Agency. An in doing so the question of acceptable exposures of
citizens who are not beneficiaries or consumers should be separated from
that which may be acceptable by consumers. The latter are free to make
their own choices regarding the use of radioactive gas and the acceptance
of the consequent radiation exposures.

(Similar disturbing questions arise in connection with other risks
and side effects of nuclear gas stimulation explosions. Many Colorado citizens who will receive no benefits are expected to accept not only the risks
from radiation exposures due to flaring, but seismic damage to their property
and to natural structures as well as the threat of radioactive ground water
contamination. These risks and effects will result in a general decline in
the quality of life in Colorado and in the reputation of our state as a desirable place to live. For these reasons it appears to us that the citizens
of Colorado should have the deciding voice on the question of the acceptability of nuclear explosive technology in our state.)

6. Recommendations: It is the view of our committee that neither the Rio
Blanco Project nor any other nuclear gas stimulation experiment should be
approved until each of the following has been carried out:

(a) That the environmental impact and public health implications
of the large scale application of nuclear gas stimulation be thoroughly reviewed in public hearings.

(b) That the NCRP, EPA, and HEW be called upon to review and recommend on permissible levels of radiation exposure to the public from the peaceful uses of nuclear explosives.

(c) That the non-nuclear alternatives to nuclear gas stimulation be given a comparable degree of federal support and consideration.

(d) That the AEC provide a final impact statement that is adequately responsive to all issues raised herein and to other questions and comments on the Rio Blanco Project that are brought out in the public hearings at the Denver Federal Center beginning on March 27, 1972.

(Prepared for the Colorado Committee for Environmental Information by E. A. Martell, H. Peter Metzger, Paul Goldan and D. M. Carmichael and duly approved by the CCEI Board of Directors.)

- - - - - - - - - -

E. A. Martell
President, CCEI
325 Norton Street
Boulder, Colorado   80303

(303) 494-9809

240

Index

Aamodt, Lee, 135
Alaska, 10
Albuquerque Journal, 54, 55
Albuquerque Tribune, 54
Allott, Gordon, 131, 136
American Gas Association, 22, 26
American Geological Society, 164
Anderson, Clinton, 54
Andrus, Dale, 162
Anspaugh, L. R., 111
Arapahoe Medical Society, 165
Argon-37, 110, 212
Argon-39, 110, 212
Aronson, Harold, 78, 141, 142, 201
Arizona, 10
Arms Control & Disarmament Agency, 11
Arraj, Alfred, 92, 96, 98, 100
Aspen City Council, 201
Aspinall, Wayne, 93
Atomic Energy Act, 1946, 34
Atomic Energy Act, amended 1954, 5, 6, 44, 98, 99
Atomic Energy Commission, 29, 33, 36, 43, 49, 50, 52, 54, 55, 64, 68, 77, 79, 81, 82, 83, 89, 91, 94, 99-101, 111-113, 118, 119, 125, 129-136, 141, 150-155, 160, 163, 165, 170, 181, 188, 193, 200-205, 212-215, 217-222
  "Agreement State," 43, 44, 193
  Code of Regulations, 44
  Commercial services, 11
  Director of Licensing, 22
  Division of Applied Technology, 7, 75, 164, 222
  Division of Biology & Medicine, 7
  Division of Military Applications, 7
  Division of International Affairs, 7
  Division of Operational Safety, 7
  Division of Peaceful Nuclear Explosives (NVOO), 7, 11, 35
    Nevada Operations Office, 35, 77, 78, 86, 170
    Project Contracting & Operations Office, 170
  Hearings, 135-140
  Nevada Test Site, 57, 62, 110, 113, 181
  San Francisco Operations Office, 50
Atomic Storage Corp., 84
Audubon Society, 139, 155
Austral Oil Co., 15, 43, 54, 68, 73, 75-77, 81, 91, 98, 107, 156, 217, 218
Australia, 10

Ball Bros. Research Corp., 88
Baruch Plan, 2
Brill, Kenneth, 76
Bronco Project, 13
Brotzman, Donald, 213
Brown, Paul N., 191
Bureau of the Budget, U.S. 11
Bureau of Land Management (Interior Dept.), 160-163

Bureau of Mines (Interior Dept.), 7, 14, 50, 52, 117, 217, 222
Butz, Earl, 152

California, 9, 10, 42
  Los Angeles Basin, 66, 67
Calvert Cliffs Coordinating Committee, 133
Calvert Cliffs, Sup. Ct. decision, 5, 132-133
Carbon-13, 110
Carbon-14, 12, 212
Cargo, David, 54
Carryall Project, 9
CER Geonuclear Corp., 13, 15, 17, 68, 75-79, 81, 85, 91, 98, 125, 128, 129, 139, 141, 142, 144, 149, 150, 151, 154-158, 160, 165, 169-171, 180, 193, 195, 197, 199, 202, 204
Cesium-137, 15, 109, 204, 213
Cheney, 191
Citizens Concerned About Rulison, 90
Citizens for Colorado's Future, 165, 191, 197
Coffer, H. F., 76
Colburn, William A., 84
Colorado, 2, 16, 44, 45, 74, 79, 89, 96, 126, 127, 130, 138, 150, 153, 172, 175, 196, 197, 199, 201, 204, 209, 210, 212, 227
  AEC contract with, 18
  Amendment #10 (Colo. referred proposal, 1974), 79, 126, 202, 205, 219, 221
  Administrative Code, 194
  Denver District Ct., 190, 192
  Dept. of Education, 227
  Dept. of Labor, 227
  Dept. of Public Health, 76, 81, 83, 85, 87, 93, 128, 154-156, 163, 165, 195, 203-205, 227
    Director of Air & Radiation Hygiene Div., 78
  Dept. of Natural Resources, 76, 79, 151, 227
    Director of Natural Resources, 78, 79, 84, 85
  Div. of Highways, 78
    Chief Engineer, 78, 81
  Geographical, 74, 175, 183, 185
    Aspen, 43, 44, 102, 113, 201
    Battlement Creek, 119
    Battlement Mesa, 74, 82
    Boulder, 88, 198, 201
    Brush, 84
    Colorado River, 110, 156, 185
    DeBeque, 141, 149, 185
    Denver, 17, 18, 21, 41, 78, 88, 119, 126, 134, 140, 161, 162, 201
    Fort Collins, 135
    Garfield County, 74, 150, 152, 185
    Glenwood Springs, 75
    Golden, 139
    Grand Junction, 74, 128, 148, 154, 199, 200
    Grand Valley, 74, 103, 185
    Harvey Gap Dam, 86
    Meeker, 13, 17, 18, 87, 129, 134-136, 140, 141, 143-149, 150, 151, 160, 185
    Mesa County, 74, 150, 152
    Moffat County, 152
    Piceance Creek Basin, 13, 17, 18, 125, 128, 141, 157, 169-171, 173, 182-185, 190, 216, 217
    Rangely, 141
    Rifle, 15, 75, 102, 145-149, 185
    Rio Blanco County, 150, 152, 185
    Silt, 86
    White River Nat'l Forest, 74
  Geological formations, 104, 184, 187
    Commerce City-Darby Geological Fault, 199

Fort Union, 170, 177, 184, 187

Green River, 184, 187

Mesaverde, 75, 82, 103, 104, 170, 177, 184, 187

Wasatch, 184, 187

Governor's Oil Shale Commission, 87

Ninth Judicial Dist., 91

Oil & Gas Conservation Commission, 77, 153, 156-158, 162, 227

Commissioners, 157

Proposed Office of Technology Assessment, 224-229

Public Utilities Commission, 45

State Attorney General, 84, 192

State Courts, 81, 91

State Legislature, 161, 203

Water Pollution Control Commission, 84, 85, 88, 153, 156, 158, 159, 163, 180, 191-195, 203, 221

Water Quality Control Commission, 203-205, 221-227

Colorado American Civil Liberties Union, 88, 91

Colorado Atomic Energy Act (proposed), 18

Colorado Committee for Environmental Information (CCEI), 88-90, 93, 137, 138, 213

Colorado Daily, 88

Colorado Interstate Gas Co. (CIG), 43, 81, 107, 156, 201, 217

Colorado Medical Society, 165

Colorado Open Space Coordinating Council, 91, 137, 139, 155, 191

Colorado Public Health Assoc., 162

Colorado School of Mines, 139, 199, 204

Mineral Resources Institute, 22, 24, 27

Potential Gas Committee, 24, 26, 139

Colorado State University, 135, 161, 165

Colorado Regional Council, 152

Continental Oil Co., 78, 204

Cook, Jerry, 191

Cooley, Frank, 87, 136

Crowther, Richard, 91

Cullen, David, 155

Defense Dept. (U.S.), 8

Delta Citizens Concerned About Radiation, 102, 198

Denver Federal Center, 94

Denver Post, 87, 97, 201, 203

"Diamond" device, 181

Dominick, Peter, 54, 165, 198

Dow Chemical Co., 154

Doyle, William E., 91

Dugan, Paul, 173, 190

Dumont, Martin G., 91

Eames, William, 91

Earthquakes, 55, 64, 118, 119, 186

El Paso Natural Gas Co., 13, 14, 17, 49, 50, 52, 68, 156

Emerson, John, 93

Energy crisis, 21, 27, 126, 164, 165, 171, 201, 211, 220

Energy Resource Development Agency Nuclear Regulatory Commission, 44, 45, 220

Engdahl, David, 191-193, 195

Environmental Action of Colorado, 137, 198

Environmental Impact Statement for the Prototype Oil Shale Leasing Program, 183

Environmental Impact Statement: Rio Blanco Gas Stimulation Project, 118, 160, 176, 184, 185, 187, 204, 212, 213, 217

Addendum, 176

Environmental Protection Agency, 2, 39, 40, 45, 161

243

Division of Criteria & Standards, 39, 40
Environmental Sciences Services Admin. (ESSA), 8
Air Research Laboratory, 83
Environmental Task Force, 191
Equity Oil Co., 17, 125, 129, 141, 144, 157, 161, 170, 190
Evans, David, 139, 155, 156, 158

Fairbanks, Richard, 152
Farmington Daily Times, 49, 54
Federal Register, 133
First National City Bank of Houston, 189
Fission, 15, 56, 57, 89, 108
Flemming, Edward, 222
Francis, E. Lee, 55
Fusion, 15, 56, 57, 109

Gasbuggy in Perspective, 61
Gasbuggy Project, 2, 13, 14-17, 34, 43, 49-69, 73, 79, 103, 125, 142, 178, 179, 181, 205, 214-216
Blast damage, 62
Chimney, 57
Economics, 57, 67, 80, 81, 92, 216, 217
Flaring, 89, 180
Fracturing, 57, 58, 61-62, 214, 217
Gas composition, 62, 63, 107-109
Gas production, 57-62, 179, 180
Nuclear device, 56, 103
Public Information & Observer Program, 52
Radioactivity, 64-67, 112, 113, 180
Seismic effects, 62-64, 86, 96
Site, 27, 55, 89, 113, 115
Gaylord, Charles, 155, 156, 191
General Accounting Office (GAO), 217
Gitchell, Ronald, 136
Gnome Project, 9

Gofman, J. W., 41
Grier, Herbert, 139, 151
Gueymard, A. C., 189

Hamburger, Richard, 7
Hampton, Clyde, 78
Hansen, Roger, 78
Hardhat Project, 61, 214
Harvey, Harold, 191
Haskell, Floyd K., 131, 153, 159, 160, 162, 164, 165, 197, 203
Hawaii, Office of Science Advisor, 227
Hoffman, Milton, 158
Hogan, Mark, 92
Holzer, A., 61, 68
Hosmer, Craig, 11
Hueckel, Glen, 218
Hydraulic fracturing, 138, 216

Idaho, 10
Interior Dept. (U.S.), 2, 8, 10, 13, 50, 79, 130, 131, 137, 150, 152, 159, 160, 162, 163
International Nuclear Test Ban Moratorium, 9
International Resources, Inc., 157
International Test Ban Treaty, 10, 13

Jacoe, P. W., 93
Japan, 2
Hiroshima, 16, 198
Jicarilla Apache Indian Tribe, 55
Johnson, Gerald W., 164
Johnson, James, 153

Keller, Glen E. Jr., 154, 163, 165
Kelly, John S., 181
Kennecott Corp., 10
Krypton-85, 12, 56, 64, 65, 66, 97, 110, 212

Lamm, Richard, 162
Lawrence Livermore Radiation Laboratory (LLL), 7, 13, 17, 50, 76, 149, 158, 170, 175, 202, 204, 214, 216, 222

244

Lewis, Howard, 163
Los Alamos Scientific Laboratory
  (LASL), 15, 77
Louisiana, 212
Love, John, 18, 76, 79, 90, 92, 96,
  127-131, 149, 150-153, 159, 162,
  198, 200, 220
  Governor's Advisory Commis-
    sion on Rio Blanco, 18, 128-
    130, 137, 141, 150-152, 190,
    226

Martell, Edward, 89, 138, 213
McGee, Gale, 164, 217
Meeker Herald, 141, 148, 151, 160
Meeker Town Council, 152
Mercury-203, 110
Metzger, H. Peter, 88, 93, 94, 137,
  138, 155, 156
Miller, David, 78
Miniata Project, 15, 181
  Nuclear device, 125, 181
Mitchell, William, 135
Montoya, Joseph, 54
Morris, Thomas, 54
Morse, Richard, 164
Morton, Rogers C. B., 131, 159,
  162
Munger, Mary Alice, 163

Nader, Ralph, 161
National Academy of Sciences, 40
  BIER Report, 40
National Center for Atmospheric
  Research (NCAR), 89, 138
National Environmental Policy
  Act (NEPA), 5, 132, 133, 159,
  162, 197, 218
National Jewish Hospital, 156
National Resources Defense
  Council, 160
National Science Foundation,
  227
  Research Applied to National
    Needs (RANN), 227
Natural gas:
  Price, 27

Resources, 25, 26
Use predictions, 22-25
Synthetic, 26
Nevada, 57, 62 204
  Las Vegas, 17, 75
New Jersey, Atlantic City, 41
New Mexico, 2, 52, 53, 83, 175
  Carlsbad, 9
  Dept. of Health & Welfare, 54
  Dulce, 55, 64
  Farmington, 13, 55
  Geological formations
    Ojo Alamo, 51, 62, 64, 65
    Pictured Cliffs, 51, 57, 62
  Navajo Dam, 5
  San Juan Basin, 13, 51, 66, 68
Nixon, Richard, 18, 152, 165, 200
Nuclear Regulatory Commission,
  2, 44, 45, 220

Oak Ridge National Laboratory,
  1, 14
Oil Shale Co., The (TOSCO), 18,
  128, 135-137, 157, 184, 188, 189
Oil shale development, 128-130,
  137, 169, 173, 182-190
Oil Shale Regional Planning
  Commission, 136, 151, 152
O'Leary, John F., 22, 26

Pauling, L., 44
People United to Reclaim the En-
  vironment (PURE), 90
Plowshare Program, 1-28, 42-45,
  49, 52, 67, 69, 76, 125, 137, 140,
  143, 161, 170, 201, 210, 212-224,
  229
Plowshare Symposiums, 7, 8, 12
Power Reactor Demonstration
  Program, 6, 11
Prouty, Dick, 201

Radiation, 11, 12, 14, 16, 37-42,
  65-67, 109-113, 222
  Federal Radiation Council
    (FRC), 38-40, 100, 111

International Commission on Radiation Standards, 38

National Commission on Radiation Protection & Measurements, 38

Radiation Protection Guide, 39

Radiation standards, 109, 111, 112

Radioactive contamination, 12, 36, 65-67, 108, 109, 156, 210, 212-214, 221

Ray, Dixie Lee,152, 159, 198, 200

Richter Scale, 96, 199

Rifle Chamber of Commerce, 152

Rifle City Council Questionnaire, 102

Rifle Telegram, 79, 95, 102, 148, 160

Rio Blanco Project, 2, 17, 18, 34, 35, 125-206, 212-217

  AEC jurisdiction, 154, 155

  Blast damage, 139, 155, 157, 158, 161, 182, 200

  Chimney, 177, 178, 182, 186, 187, 213, 216

  Contract, 149, 150, 158, 169, 170, 193

  "D+30 Day Report," 199

  Economics, 138, 139, 155, 156, 170, 171, 173-175, 181, 182

  Employment, 141, 189, 216

  Feasibility Study, 169

  Flaring, 139, 141, 196, 220

  Fracturing, 177, 178, 186-188

  Gas composition, 180, 199

  Gas production, 173, 174, 178-190, 199, 216

  Governor's Advisory Commission, 18, 128-130, 137, 141, 150-152, 190, 226

  Ground waters, 138, 139, 141, 158, 159, 196, 200

  Litigation, 190-197

  Objectives, 170-176

  Nuclear devices, 177, 178, 181, 182, 198

  Phase I, 126, 129, 136, 150, 152, 153, 162, 163, 170, 171, 176, 182, 186, 189, 190

  Phase II, 173, 174

  Phase III, 174

  Phase IV, 174

  Project Definition Contract, 170

  Public opinion survey, 143-149

  Radiation standards, 154, 155

  Radioactivity, 130, 138, 139, 153, 155, 156, 158, 159, 161, 180, 181, 196, 199

  Safety measures, 155, 182, 196

  Seismic effects, 130, 139, 155, 161, 182, 186-188, 199, 200

  Site, 27, 153, 170-172, 188

  Technology, 155, 169, 176-182

  Taxes, 141, 142

  Water disposal, 180

Rio Blanco County Planning Commission, 129, 136, 141, 142

Robinson, W. L., 111

Rocky Mountain Arsenal, 119

Rocky Mountain Center on the Environment, 78

Rocky Mountain Gas Co., 43, 113

Rocky Mountain News, 89, 201

Rold, John, 93, 139, 140

Rulison Project, 2, 16, 34, 35, 42-45, 64, 68, 69-120, 126, 127, 139, 148, 165, 179, 181, 190, 198, 205, 215

  Atmospheric conditions, 87

  Blast damage, 96, 115-119, 200

  Chimney, 103

  Contract, 79

  Economics, 80, 81, 82, 106, 107, 216-218

  Feasibility Study, 75

  Flaring, 89, 97, 100-102, 111, 112, 180, 220

  Fracturing, 103, 105, 217

  Gas composition, 107-109

Gas production, 106, 107, 179
Ground waters, 83, 85, 89, 109, 110, 113, 119
Joint Office of Information, 79, 81-83, 90
Litigation, 91, 92, 94-101, 178
Objectives, 15, 80
Nuclear device, 79, 82, 89, 103
Project Definition Report, 77, 78
Project Rulison Objectives & Experimental Details, 79
Radioactivity, 93, 109-114, 180
Seismic effects, 89, 115-117, 119
Safety measures, 82, 87, 92, 100, 101
Technical results, 103-120
Santa Fe Railroad, 9
Santo, Henry, 190, 198
Schlesinger, James, 7, 17, 151
Schoeberlein, William, 192
Schroeder, Patricia, 164, 201
Seaborg, Glen T., 91, 97, 131, 136, 151
Shell Oil Co., 128
Signal Drilling Co. (Div. of Superior Drilling Co.), 78
Sloop Project, 10
Smith, James III, 91
Smith, Sam, 14
Soviet Union, 2
State Dept. (U.S.), 11
Stern, Carlos, 161
Sternglass, Ernest, 93
Storer, J., 41
Strontium-90, 12, 15, 109, 204, 205, 213
Tamplin, A. R., 41
Technology, natural gas
  Acid leaching, 27, 28
  Conventional explosive, 28
  Hydraulic fracturing, 28, 29
  Nuclear stimulation, 29-37
Teller, Edward, 8, 32, 201
Ten Eyck, Thomas, 78, 84-86, 151, 205

Terrill, James, 12
Texas, Houston, 73, 189
TNT, 16, 73, 198
Toman, John, 149, 216
Transportation Dept. (U.S.), 45
Travois Project, 10
Tri-State Fossil Fuels Energy Conference, 216
Tritium, 12, 13, 57, 64-68, 97, 109, 110, 112, 114, 181, 203, 204, 212, 213
Truman, Harry S., 3
Tucker, William, 192

United Nations, 2
  United Nations Scientific Commission on Effects of Radiation, 38
United Press International, 73
United States:
  Army Corps of Engineers, 10
  Coast & Geodetic Survey, 8
  Congress, 2, 3, 7, 11, 16
    Fourth Cong. Dist. (Colo.), 153
    House Committee on Interior, 93
    House of Representatives, 201
    House Subcommittee on Mines, 222
    Joint Committee on Atomic Energy, 3, 11-14, 42, 49
    Office of Technology Assessment (OTA), 227
    Senate Interior Subcommittee on Public Lands, 162, 164
    Senate Subcommittee on Public Works, 131
  Federal Aviation Authority, 198
  Federal Code of Regulations, 43
  Federal Court:

District Ct., Denver, 95-97
Rulison litigation, 91-92, 94, 101
Tenth Circuit Ct., 94
Federal Power Commission, 22, 26, 27, 45
Geological Survey (Interior Dept.), 8, 109, 110, 160
Unit Development Agreement, 76, 77
Unit Operations Agreement, 76, 77, 157, 158
Rio Blanco Federal Oil & Gas Unit, 170
Public Health Service, 7, 14, 83, 109
Weather Bureau, 51
University of Colorado, 88, 143
Student Anti Pollution Committee, 88, 95
University of Connecticut, 161
University of Denver, 78, 165
Utah, 175, 183
Office of Science Advisor, 227
Salt Lake City, 17, 169

Vanderhoof, John, 200
Van Poolen, H. K. & Assoc., 179, 204
Verdieck, Emma, 161
Vigil, Charlie, 55

Wagon Wheel Project, 16, 17, 68, 175
Weinberg, Alvin, 1
Weiner, Robert, 78
Weickman, Ben, 188-190
"Well bore effect," 32, 61
Werth, G. C., 61
Western Slope Gas Co., 113
Williams, Robert H., 89, 93
Winter Olympics, 1976, 191
Wolf Ridge Minerals Co., 128
Wright, Skelly, 133
Wyoming, 16, 17, 164, 175, 183
Pinedale, 82

Yardumian, Louis, 137
Zelle, Max, 135